RE[illegible]D

Upgrade your Buiness DNA with AI

SUSAN FREW, C.S.P.

DISCLAIMER: This book provides general business guidance and is not a substitute for professional financial, legal, or technical advice. Results may vary. Tool recommendations are current as of January 2025 and subject to change. Case studies have been modified to protect client confidentiality.

Published by Susan Frew Speaks

ISBN: [To be added]

First Edition, 2025

DEDICATION

To the knowledge workers who refuse to be replaced.

You're the analysts who see patterns others miss. The strategists who turn data into direction. The relationship builders who know that connection can't be coded. You understand what the hype merchants don't: AI isn't here to take your job. It's here to make you unstoppable.

This book is for you. The human in the loop who's ready to become the best version of yourself, amplified by technology that finally works the way your brain does.

And to my fellow neurodivergent thinkers, the ones whose brains move faster than linear systems ever could: the ADHD strategists who see twenty possibilities at once, the OCD perfectionists who catch what others overlook, the dyscalculic problem-solvers who learned to think in patterns instead of numbers. AI doesn't just level the playing field for us. It tilts it in our favor.

Our neuro-spicy has been waiting for a time like this.
Your different is your advantage. Let's make it count.

I wouldn't be here without the people who saw me before I could see myself. William, you believed in me long before I did, and you gave me the space and support to chase this dream when it felt impossible. To my Polish relatives who loved me from birth and continue to show up with such warmth, thank you for being part of my story from the very beginning. To my new Italian family in New Jersey, your gracious acceptance means more than you know. Finding you has been like finding missing pieces of myself. Meeting people who actually look like me unlocked something profound. My DNA always knew who I really was, but revealing it brought a whole new level of confidence and empowerment I never expected. Understanding my roots helped me understand myself, and that foundation changed everything.

Foreword

On behalf of the National Association of Flour Distributors, it is our pleasure to welcome you to this special edition of *RECODED: Upgrade Your Business DNA with AI.*

As an industry rooted in tradition yet continually shaped by innovation, we find ourselves at a pivotal moment. The forces transforming how we operate, compete, and serve our customers are accelerating and none more so than artificial intelligence. What was once a distant concept is now a practical, powerful tool that is redefining the future of business including our own.

This year's Annual Convention at the Arizona Biltmore is centered on equipping our members with the insights and tools needed to navigate this evolving landscape. Susan Frew's work stands at the intersection of strategy, technology, and real-world application. Her ability to translate complex AI concepts into actionable business practices makes this book not only timely but also essential.

This book is intended to extend the impact of her workshop beyond the stage and providing you with a resource you can return to as you begin to explore and implement AI within your own organizations. Whether you are just starting your journey or looking to deepen your understanding, we believe this book will serve as both a guide and an inspiration for what is possible.

We encourage you to read it with curiosity, share it with your teams, and most importantly, use it as a catalyst for thoughtful action.

We are grateful to Susan for partnering with our Association and for tailoring her insights specifically for our community. Her contribution helps ensure that our members are not only prepared for the future but positioned to lead it.

With appreciation and a shared commitment to what lies ahead,
David Scruggs, Jr., 2026 Convention Chair
National Association of Flour Distributors

NAFD Board of Directors:
Ashley Koerner, Chairman of the Board; Eric Metzendorf, President; R. David Scruggs, Jr. 1st Vice President; Rich Wise, 2nd Vice President; Kevin Johnson, Secretary/Treasurer; Immediately Past Presidents: Steve Tardella, Nicholas DePalma, Ashley Koerner; Directors: Uri Alcalay; Duane Bruinsma; Sheena Dau; Theodore (Ted) Heim, Jr.; Jeremy Heim; Pamella Palmer; Erin M. Ruhl; John D. Traynor; and Phil Zilka; Allied Representative John Johanson; Miller Representative David Rowe.

Geraldina (Gerri) Cristantiello, Executive Director

AUTHOR'S NOTE

I need to tell you something that might surprise you: I almost didn't write this book.

Not because I don't have the expertise. I've spent 3,000+ hours training in AI, run multimillion-dollar companies, and taught thousands of business leaders how to transform their operations with technology. I've been an international VP for a Fortune 20 tech company, an INC 5000 CEO, and a Google AI Certified keynote speaker who shows audiences exactly how to compete on awesome with AI.

This book exists because you shouldn't have to learn these lessons the way I did.

If you're reading this, you're likely already successful. You've built something. You lead people. You make decisions that matter. But you're also standing at a crossroads, wondering if AI is going to make you obsolete or if there's a way to make it your unfair advantage.

I'm here to tell you: there is. And it doesn't require you to become a data scientist, hire a team of developers, or abandon everything you've built. It requires you to recode your business DNA, one workflow at a time.

Let's get to work.

CONTENTS

PREFACE

Why This Book, Why Now

The AI conversation has gotten ridiculous.

On one side, you've got the doomsayers predicting mass unemployment and robot overlords. On the other, you've got tech bros promising AI will solve world hunger while they're still figuring out how to make it stop hallucinating about basic math.

Meanwhile, actual business owners and executives are stuck in the middle, trying to figure out what any of this means for the companies they run and the people they employ.

Here's what it means: AI is the biggest operational shift since the internet, and if you don't learn to use it strategically, someone who does will eat your lunch.

But here's what it doesn't mean: firing your team, replacing human judgment with algorithms, or turning your business into a soulless tech experiment.

I wrote this book because the conversation about AI in business has been dominated by two groups who have no business leading it. The first group consists of technologists who've never run a P&L or managed a team through a tough quarter. The second group consists of consultants who talk about AI transformation in theoretical frameworks that sound impressive in boardrooms and fail spectacularly in execution.

I'm neither. I'm a business operator who learned AI because I had to compete, not because I wanted to impress anyone with buzzwords.

This book is different because it comes from that perspective. It's not theoretical. It's not hype. It's not written by someone who's never had to fire a vendor, negotiate a contract, or explain to a board why revenue missed projections.

It's written by someone who has done all of that and discovered that AI, when used strategically, changes everything.

What Makes This Book Different

Most AI business books fall into one of three categories:

The Technical Manual. Written by engineers for engineers, full of code samples and neural network architectures that mean nothing to the people actually running businesses.

The Fear-Mongering Manifesto. Designed to convince you that AI is coming for your job, your industry, and possibly your soul unless you panic-buy whatever solution they're selling.

The Utopian Fantasy. Promises that AI will revolutionize your business overnight without addressing the messy reality of implementation, resistance, or the fact that most companies can't even get their email marketing automated correctly.

This book is none of those.

This is a practical guide for business leaders who want to transform their operations using AI without firing their teams, abandoning their values, or pretending they suddenly became data scientists.

It's organized around a framework I call Business DNA: the combination of History, Habits, Tech Stack, and People that defines how your organization operates. You'll learn how to systematically recode each element using AI workflows that are proven, practical, and designed for real humans doing real work.

You won't find theoretical discussions about artificial general intelligence or philosophical debates about consciousness. You will find step-by-step playbooks for automating your finance workflows, transforming your marketing operations, and freeing your best people from repetitive tasks so they can do the strategic work that actually moves your business forward.

Who This Book Is For

This book is for business leaders who are tired of being told that AI is either going to save the world or destroy it, and who want practical answers about what it can do for their organizations right now.

You're an executive, entrepreneur, or business owner who understands that technology is a tool, not a religion. You care about results, not hype. You want to stay competitive without sacrificing the people and culture that make your business worth running in the first place.

You might be skeptical. Good. Skepticism means you're thinking critically instead of swallowing marketing pitches whole.

You might be overwhelmed. Also good. That means you understand the magnitude of what's happening and you're smart enough to know you need a plan.

You might even be a little scared. That's human. But here's what you need to know: the businesses that thrive in the age of AI won't be the ones that had the biggest budgets or the fanciest technology. They'll be the ones that moved decisively while others were still debating whether to start.

How to Use This Book

This isn't a book you read once and put on a shelf to collect dust next to all the other business books you swore you'd implement but never did.

This is a working manual.

It's organized into five parts that walk you through a complete transformation process, from understanding your business DNA to implementing your first AI workflows to scaling across your entire operation.

You can read it straight through to understand the complete framework, or you can jump directly to the workflow chapters that address your most pressing needs right now. Finance keeping you up at night? Start with Chapter 9. Marketing eating your budget without delivering results? Chapter 11 has your playbook.

This book includes:

- Real-world case studies from companies that have successfully implemented these workflows
- Step-by-step implementation guides with realistic timelines and resource requirements
- Templates and frameworks you can adapt to your specific business
- Resistance strategies for handling the inevitable pushback from your team
- Measurement frameworks so you know exactly what's working and what needs adjustment

The goal isn't to turn you into an AI expert. The goal is to give you the tools, frameworks, and confidence to transform your business operations strategically and systematically.

A Word About Fear

Let's address the elephant in the room: you're probably afraid AI is going to replace you or your team.

I get it. The headlines are terrifying. The predictions are dire. And if you listen to certain corners of the internet, we're all going to be obsolete by next Tuesday.

Here's the truth: AI is going to replace tasks, not people. And the businesses that thrive will be the ones that free their people from soul-crushing repetitive work so they can focus on the strategic, creative, and relationship-driven work that actually creates value.

The CFO who spends 60% of their time on data entry isn't being replaced by AI. They're being freed to do the analysis, forecasting, and strategic planning that makes them invaluable. The marketing manager drowning in content creation busywork isn't losing their job. They're gaining the capacity to focus on strategy, messaging, and campaign optimization.

This book shows you how to make that happen in your business. Not someday. Not when the technology gets better. Now.

Let's recode your business DNA and build something extraordinary.

INTRODUCTION

Your Business Has DNA

Every business has DNA.

Not the biological kind, obviously. But a code nonetheless. A set of instructions that determines how your organization operates, adapts, and survives in competitive markets.

Your business DNA consists of four elements:

History. The decisions, experiences, and lessons that shaped how you operate today.

Habits. The workflows, processes, and repeated actions that define your daily operations.

Tech Stack. The tools, systems, and platforms that enable your work.

People. The skills, personalities, and capabilities of your team.

Together, these four elements create your organizational identity. They determine how quickly you move, how effectively you compete, and whether you thrive or merely survive when markets shift.

Here's the problem: most businesses are running on legacy DNA. Code written years ago, when the competitive landscape looked completely different. When customer expectations were lower. When margins were fatter. When you could succeed by simply showing up and doing decent work.

That world is gone.

The businesses winning today aren't necessarily smarter or better funded. They're the ones that rewrote their code. They recoded their business DNA to operate faster, leaner, and more strategically than their competitors. And the primary tool they used wasn't more people, bigger budgets, or longer hours.

It was AI.

The Wake-Up Call Nobody Wants

Let me tell you about a conversation I had with a CEO last year. We'll call him David.

David runs a successful professional services firm. Twenty years in business. Solid reputation. Loyal clients. Good margins. By any traditional measure, his business was healthy.

Except it wasn't growing. Revenue had plateaued. His best people were maxed out. And every time he looked at hiring additional staff, the math didn't work. The margins got thinner. The complexity increased. The leverage disappeared.

"I'm working harder than I did ten years ago," he told me, "and making less money per hour of effort. Something's broken, but I can't figure out what."

Here's what was broken: David's business DNA was optimized for a market that no longer existed.

His workflows were built around manual processes that made sense fifteen years ago but created bottlenecks now. His tech stack consisted of tools purchased at different times for different reasons, none of which talked to each other. His best people spent 70% of their time on administrative tasks and 30% on the high-value strategic work that actually generated revenue.

And his competitors? The ones who were growing while he was stuck? They had recoded. They were using AI to automate the bottlenecks, streamline the workflows, and free their people to focus on the work that mattered.

David wasn't failing because he was lazy or incompetent. He was stuck because he was still operating with DNA written for a different era.

Sound familiar?

The AI Advantage Nobody Talks About

Here's what most people get wrong about AI in business: they think it's about automation.

It's not. At least, not primarily.
AI is about leverage.

It's about taking your best people and multiplying their capacity to create value. It's about eliminating the repetitive, soul-crushing work that turns talented professionals into glorified data processors. It's about freeing your team to do the strategic, creative, relationship-driven work that can't be automated.

When you recode your business DNA with AI, you don't replace people. You create what I call the Human Dividend: the additional capacity and capability that emerges when skilled professionals stop spending their time on tasks that machines can handle better.

Consider what happens when you implement AI workflows in your finance operations. Your CFO isn't replaced. They're freed from spending 60% of their time on data entry, reconciliation, and report generation. That time gets reallocated to analysis, forecasting, and strategic planning. The work that actually moves the business forward.

The value isn't that you eliminated tasks. The value is that you elevated your people.

That's the AI advantage nobody talks about. Not replacement. Elevation.

Why Most AI Initiatives Fail

Before we go further, let's address why most businesses screw this up.

It's not because AI doesn't work. It's because they approach implementation backwards.

They start with the technology instead of the workflow. They buy AI tools before they understand their processes. They implement solutions before they've mapped their problems.

And then they wonder why adoption stalls, resistance builds, and the expensive new AI platform sits unused while everyone goes back to the old way of doing things.

I've seen this pattern repeatedly. A company spends six figures on an AI solution. They announce it with great fanfare. They run a training session or two. And six months later, adoption is at 15% and the executives are quietly wondering if they got scammed.

They didn't get scammed. They got the sequence wrong.

Successful AI implementation doesn't start with technology. It starts with understanding your business DNA. Mapping your workflows. Identifying the bottlenecks and inefficiencies that are costing you money, time, and competitive advantage. And then, only then, selecting the AI tools that solve those specific problems.

This book gives you that sequence. The right way to transform your operations without wasting money, time, or organizational goodwill on initiatives that fail.

What You're About to Learn

This book is organized into five parts that take you through a complete transformation process:

PART I: Understanding Your Business DNA establishes the foundation. You'll learn how to map your organizational code and identify the elements that need transformation.

PART II: The Recode Process introduces the three-phase framework for implementing AI: Pilot, Measure, Scale. This is the systematic approach that prevents failure and ensures sustainable change.

PART III: Foundation Workflows gives you detailed playbooks for the three core business functions that every organization needs to transform: Finance, Operations, and Marketing.

PART IV: Integration and Scale shows you how to expand AI implementation across your organization, handle resistance, and build the infrastructure for sustainable transformation.

PART V: The Future of Your Recoded Business addresses the emerging trends, ethical considerations, and strategic decisions that will define your competitive advantage going forward.

Each section builds on the previous one, creating a complete framework for transformation. But you don't have to read it sequentially. If you're facing an immediate challenge in marketing, jump to Chapter 11. If your finance operations are the bottleneck, start with Chapter 9.

The goal is implementation, not theory. Use this book as a working manual, not a manifesto.

Your Competitive Moment

Here's what you need to understand: we're in a unique moment in business history.

AI is powerful enough to create genuine competitive advantage but not yet ubiquitous enough that everyone is using it well. That window won't last forever.

The businesses that move now, that implement strategically and systematically, will build advantages that compound over time. Better workflows mean faster execution. Faster execution means more market share. More market share means more data to train better AI. Better AI means even more efficient workflows.

It's a flywheel. And the businesses that build it first will dominate their markets for the next decade.

The businesses that wait? They'll be competing on price because they won't be able to compete on speed, quality, or innovation.

You get to choose which category you're in.

This book gives you the tools to choose wisely. To recode your business DNA. To build the competitive advantage that makes you not just relevant but dominant in your market.

Let's GO!

What You'll Learn and How to Use This Book

Most AI books tell you what's possible. This one shows you exactly how to do it. This book is a practical field guide. By the end, you'll be able to:

Map your business DNA (history, habits, tech stack, people) and spot the highest-leverage fixes.

Design safe, human-centered AI workflows with clear policies, data protections, and review gates.

Run a 30-day pilot with one owner, one metric, and a simple dashboard, then decide: kill, keep, or iterate.

Automate the routine, humanize the exceptions so your team spends more time on the moments that matter.

Measure what matters (cycle time, exception rate, cost per document, adoption) and translate wins into budget.

Coach past resistance, including the Digital Denialists, with scripts, incentives, and small wins.

Who This Is For

Leaders, owners, and operators who want results in 90 days without turning their culture into a robot factory. If you can influence process, people, or tools, you're in the right place.

The 90-Day Deployment Plan (Step by Step)

Use this as your scaffolding. Every chapter links to a worksheet or checklist you can run Monday morning.

Month 1: Discover & Design (Weeks 1-4)

Map your mess, pick your battles, baseline your metrics.

Go old school to go new school. Get everyone around the table to map everything out. Whiteboards. Sticky notes. Paper.

AI TOOL: (process mapping or transcription assistant)

PROMPT IDEA: "Summarize this whiteboard photo into steps, owners, and handoffs. Flag bottlenecks."

Pick three Monday-ready workflows: high volume, rules-based, measurable.

Design guardrails: acceptable use, vendor checklist, PII rules, review gates.

AI TOOL: (policy draft assistant)

PROMPT IDEA: "Draft an AI Acceptable Use policy for a 50-person services firm. Keep it one page, add examples."

Baseline your metrics: cycle time, exception rate, cost per document, error rate.

PROMPT IDEA: "Create a simple KPI sheet with definitions and how to pull each metric weekly."

Month 2: Pilot (Weeks 5–8)

Launch small, measure tight, prove value fast.

Write a 30-Day Pilot Charter: one owner, one metric, success criteria, risks, communications plan.

> **AI TOOL:** (document generator / checklist builder)
>
> **PROMPT IDEA:** "Turn this pilot outline into a one-page charter with scope, RACI, timeline, and a risks table."

Launch with humans in the loop. Reviewers approve exceptions. Log edge cases.

Guardrail: No customer-facing changes without human review.

Measure weekly. Show a tiny dashboard and a five-minute demo of wins.

> **PROMPT IDEA:** "Generate a weekly pilot update for execs with three bullets: what shipped, metric movement, next risk."

Month 3: Rollout (Weeks 9–12)

Scale what works, kill what doesn't, reinvest the savings.

Kill, keep, or iterate based on the metric.

Train and transition: role redesign, incentives, job aids, and office hours.

> **AI TOOL:** (training module builder)
>
> **PROMPT IDEA:** "Convert this SOP into a 10-minute lesson with three quiz questions and a readiness checklist."

Set up governance and budget: portfolio of workflows, sunset junk tools, reinvest savings.

> **PROMPT IDEA:** "Summarize pilot savings and propose a 12-month roadmap with three new workflows and budget."

How We'll Teach Prompting (Lightbulb Callouts)

Throughout the book you'll see lightbulb callouts you can copy and paste:

> AI Tool: A specific tool category you can slot in ().
> Prompt Idea: A ready-to-use prompt, tuned for the task.
> Guardrail Tip: A quick risk to avoid.
> Metric to Watch: The one number that proves value.

You'll also get a Prompt Recipe at the end of key chapters (finance, ops, marketing, agents) so you can reuse and adapt.

What This Book Is Not

Not hype about model names. We focus on capabilities and outcomes.
Not theory without tools. Every chapter ends with a 10-minute action item.
Not tech for tech's sake. If a workflow doesn't move a metric, it doesn't make the cut.

Breadcrumbs (If You Want Help)

I work with a small number of teams on 90-day frontier plans, onsite workshops, and pilots.

Use case audit: Pick three Monday-ready workflows.
Pilot in a box: Charter, guardrails, success metric, dashboard.
Leader enablement: Scripts, incentives, job redesign.

Interested in a guided rollout?

Your 10-Minute Workflow Audit

List 10 recurring tasks in your team.

Circle three that are rules-based and measurable.

For each, jot the actor, inputs, decision, outputs.

Add one 💡 Prompt Idea you could try tomorrow.

Done when: You've picked one workflow to take into Month 1, Step 1.

PART I
The DNA Way of Thinking

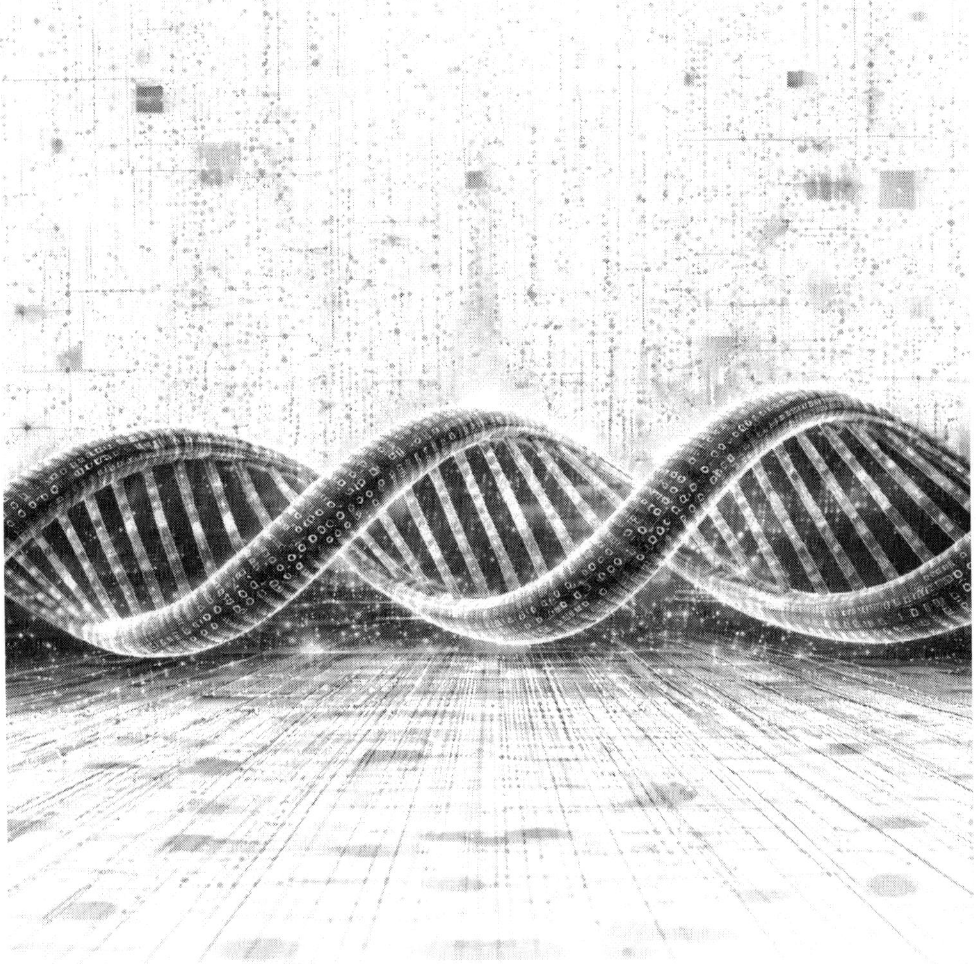

CHAPTER 1

The Day My DNA Changed

January 21, 2021. The day I found out I'd been lied to my entire life.

I'm sitting at my kitchen table. Coffee's getting cold. And I'm staring at this DNA kit I ordered weeks ago because, honestly, who doesn't want to know more about their ancestry? I thought I'd see Polish last names. My parents always told me I was half Polish. That's what I grew up believing.

I open the results.

Not a single Polish surname. Not one.

Instead? Irish. English. Italian. And then this number fills my screen: fifty-two percent Italian.

I'm staring at it thinking, "This has to be wrong." But it's not wrong. It's just the truth nobody told me.

And here's the part that really got me. I grew up in New Jersey. Do you know how many Italian families are in New Jersey? A lot. So my brain goes straight to the worst possible place: Oh my God, did I ever date any of these people?

That moment didn't just shake me up. It fundamentally changed how I see myself. For most of my life, I felt like something was off. Like I was pretending to be someone I wasn't. Turns out? I actually was.

But here's what I didn't expect. This whole experience completely changed how I think about running a business.

Your business has DNA too. It's built from your education, your experiences, your employees, your clients. Every story. Every failure you survived. Every win you celebrated. Right now, AI is reshaping how we all work, and honestly? We're being asked to recode our business DNA all over again.

For a lot of people, that's going to feel impossible. Like staring at a screen full of unfamiliar names and not recognizing your own life.

But for those who can do it? This will be the best thing that ever happened to their careers.

There's something else I need to say here, because it explains a lot about how I think, how I work, and why AI became such a powerful tool for me.

Over my lifetime, I've had ten concussions and nineteen broken bones. Along the way, that history amplified things I was already wired with: ADHD, a touch of OCD, and Dyscalculia (Dyslexia) with numbers. My brain is excellent at pattern recognition, big-picture thinking, and creative problem-solving. It is far less reliable when it comes to memory, recall, and numerical precision.

For a long time, I treated that as a personal flaw I needed to work around:or hide. I built compensating systems. I overprepared. I double-checked everything. I exhausted myself trying to operate like someone I wasn't.

Here's what I've learned since: ADHD isn't a weakness. It's a superpower:when it's supported properly. The problem isn't the wiring. The problem is forcing that wiring to operate without the right tools.

That's where AI changed everything for me.

AI doesn't replace my thinking. It supports it. It acts as an external memory. A recall engine. A second set of eyes for numbers and structure. It lets me stay in my zone of strength:strategy, synthesis, leadership:without being penalized for the cognitive areas that cost me the most energy.

I share this because many of the most capable leaders I know carry invisible limitations of their own. Neurodivergence. Brain injuries. Burnout. Cognitive load from years of responsibility. AI isn't about making people faster robots. It's about designing systems that let humans operate at their best.

That belief sits at the core of everything in this book.

What Business DNA Actually Means

Every organization has its own DNA. And I mean that literally, not as some fancy metaphor.

Your history. Your habits. Your tech stack. Your people. These combine like actual genetic building blocks to shape how work gets done and how decisions get made. You can't change what's not working until you understand what you're actually dealing with.

There's this article from Strategy+Business back in 2003. They compared corporate DNA to four key bases: structure, decision rights, motivators, and information. How these interact determines how your organization behaves. Other research says it's culture, structure, and processes.

Both are right. It's all of it.

For this book, I'm breaking it down into four parts because that's what works when you're actually doing this in real life:

- History – all the decisions, culture, and legacy norms shaping what you do today
- Habits – the routines and informal practices nobody wrote down but everyone follows
- Tech Stack – the tools and systems either helping you work or holding you hostage
- People – the skills, roles, and relationships actually making things happen

Here's one of my core principles. I say it on stage all the time, and I've proven it running actual businesses: if you want to learn something new, you have to go back to what's old first.

That means getting everyone in a room. Whiteboards. Sticky notes. Sometimes just paper and Sharpies. You need to see the whole picture before you can use AI to recode anything. It's not about the tools. It's about getting clear on what you're actually doing and why you're doing it that way.

The Four Parts of Business DNA

History

This is all the stories, decisions, and cultural beliefs piled up over the years. The stuff nobody talks about but everyone knows. These unspoken narratives shape how your employees think about risk, about change, about what success even means.

As leaders, we've got to drag these stories out into the open. Look at them. Decide which ones to keep and which ones to retire.

Because sometimes the story you've been telling yourself about your company isn't actually true anymore.

Habits

Habits are the informal rules guiding daily behavior.

Your policy manual says one thing. Your people are doing something completely different because they figured out a workaround three years ago and it stuck.

When you identify these habits, you'll see friction. Redundancy. Inefficiency you didn't even know existed. You'll also see where tiny shifts could free up massive amounts of energy for work that actually matters.

Tech Stack

Your tech stack is every tool, platform, and system you're using to get work done. When it's integrated well? It accelerates everything. Collaboration. Insights. Speed.

When it's broken? You get bottlenecks. You get shadow IT where people are secretly using tools you don't even know about because your official systems are so bad.

I've seen companies running on systems from 2008 wondering why they can't compete in 2025. The answer's pretty obvious when you map it out.

Mapping your stack shows you exactly where legacy technology is keeping your team locked in outdated patterns. It also shows where new tools could automate the repetitive garbage you're doing every day.

People

People are what bring DNA to life.

Skills. Roles. Relationships. These determine how projects actually launch. How ideas spread through your organization. How things get done when nobody's looking.

Effective collaboration needs trust. Clearly defined roles. Diverse perspectives. And here's something most people miss: you need to look at both the official org chart and the unofficial networks that actually run things.

Because the org chart? That's what you tell the board. The unofficial networks? That's reality.

Five Ways to Map Your DNA (That Actually Work)

I've been doing this work for years. These are the five practices that consistently work, along with the pitfalls I see people fall into and the metrics you should track.

1. Hold Cross-Functional Discovery Sessions

Get people from different departments in the same room. Map how work flows from start to finish. Use whiteboards or process-mapping tools to visualize each step and spot bottlenecks.

These sessions help teams share knowledge they didn't know the other team needed. They surface assumptions nobody realized they were making.

Watch out for: Leaving out frontline workers. Letting the loudest person dominate. Turning it into a blame session instead of actually learning something.

Track this: Number of cross-functional insights you capture. Handoff issues. Process pain points. That kind of thing.

2. Validate Your Assumptions with Actual Data

Use real data to identify your highest-impact areas. Cycle times. Error rates. Support tickets. Process mining tools will show you exactly where tasks are stalling or repeating for no good reason.

Watch out for: Only listening to anecdotes. Ignoring data that contradicts what leadership thinks is happening. Collecting data without any context so it's meaningless.

Track this: How much your cycle time or error rate drops after you make changes.

3. Figure Out What's Routine and What's Not

Sort your tasks into two buckets. Routine work that's repeatable and rules-based. Exception work that's unique and needs human judgment.

Automate or streamline the routine stuff. Free your people up to focus on the exceptions that actually require thinking.

Watch out for: Calling high-touch work "routine" when it's not. Automating too fast without testing. Underestimating how hard it is for people to switch back and forth between routine and exception work.

Track this: What percentage of your routine tasks you've actually automated or streamlined.

4. Co-Create Solutions with Your Employees

Get the people closest to the work involved in designing the improvements.

This isn't rocket science. Co-creation builds buy-in. It surfaces what's actually possible versus what leadership thinks should be possible. It creates a culture where people keep improving things instead of waiting to be told.

Watch out for: Treating this like a one-time checkbox exercise. Asking for suggestions and then doing nothing with them. Assuming you know better without actually listening.

Track this: What percentage of your employees are participating in improvement initiatives.

5. Build a Living Process Library

Store your documentation somewhere central where anyone can see it and update it. Include who owns what, how often it happens, what goes in, what comes out. Link to templates or tools.

Make it easy for teams to suggest updates and see progress.

Watch out for: Letting documentation go stale. Making people jump through approval hoops just to access it. Failing to write down the tacit knowledge people carry around in their heads.

Track this: How many processes you've documented and updated in the last three months.

Map Your Organization's DNA

Here's a worksheet to get you started. Check the boxes. Answer the questions. Describe what success would actually look like.

When you're done: You've identified one old story to challenge, one habit to shift, one tool to test, and one relationship to strengthen.

Things to Remember

Start here: Hold listening sessions with all your teams. Uncover the habits nobody talks about.

Don't do this: Confuse tools with transformation. A new app won't fix a broken culture. I've watched companies try. It doesn't work.

Quick exercise: List three things you do the same way every day. Which one could be automated?

Real Example: How Mapping Saved $50,000

I worked with this mid-sized manufacturing company. They were drowning in slow approvals for everything.

We mapped their DNA. Found out they had unclear decision rights. Five different managers had to sign off on every purchase. Every. Single. One.

They revised their approval thresholds. Documented their processes. Cut purchase cycle time from 14 days to 3. Saved $50,000 in administrative costs.

You're probably thinking your situation is different. More complex. It's not.

Small structural shifts unlock massive gains.

Your Turn

Stop for a second. Think about your company's DNA right now.

How are history, habits, tech stack, and people shaping your culture? Write it down. What are you actually seeing? Not what you wish you were seeing.

Use that to figure out which experiments to run first.

The Truth About DNA

Understanding DNA requires facing the truth. Doesn't matter if it's biological or organizational.

When I found out who my real ancestors were, it wasn't just about heritage. It was about identity. Purpose. What's actually real versus what I'd been told.

Same thing applies to running a business.

Companies tell themselves stories constantly. "We're customer-first." "We're innovators." "We're like family."

But when you examine the company's DNA? Those stories usually fall apart pretty fast.

A DNA test reveals genetic truth. An operational audit exposes the real culture hiding under all that branding.

Here's what I've learned: knowing your company's DNA takes guts. Real guts. It means being honest about what's working, what's not, and what's completely outdated.

The moment you face that truth? You get more powerful. You can change it. You can recode it.

But you have to look at it first.

The Emotional Side of Change

Discovery feels amazing. It can also completely disorient you.

When I saw those unfamiliar Italian names on my screen, I felt unmoored. Lost. Like the ground shifted under my feet and I didn't know where I was standing anymore.

Teams feel the same way when the business they thought they knew starts to change. Fear comes up. Resistance builds. People grab onto what's familiar because it feels safe, even when it's obviously broken.

That's where leadership matters.

Real leadership acknowledges fear. It doesn't dismiss it or pretend it's not there. Change management isn't about forcing people to comply. It's about creating belonging inside the transformation process.

When your team sees you're recoding the DNA, not replacing the people, they start to trust what you're doing.

Why Organizations Work Like Bodies

Every business has structure. Think about it like a body.

The bones are your frameworks. The muscles are your processes. The blood vessels are how information flows through the organization.

When one part weakens, the whole thing suffers.

AI doesn't replace the body. It strengthens what's already there. It's the catalyst that brings clarity to systems you couldn't see before.

Look, in 2025, AI isn't some future skill we'll need eventually. It's foundational right now. Just like reading and math are foundational.

Understanding how AI impacts decision-making? That's going to be the same level of basic skill.

Companies resisting this shift are going to end up like doctors before DNA sequencing existed. Guessing instead of diagnosing. And that's not a great place to be.

A Real Story About Trust

I consulted with a regional service company a while back. Slow workflows. Constant turnover. Outdated systems. Their processes weren't documented anywhere. Morale was terrible.

We mapped their DNA and found something crucial.

Their biggest problem wasn't tools. It was trust.

Employees didn't believe leadership would follow through on promises to improve things. They'd heard it all before. New initiatives. Big announcements. Nothing ever changed.

So we went old school. Got everyone in the same room. Covered the walls with butcher paper. Handed out markers and sticky notes. Mapped every single step of their workflows together.

The energy in that room completely shifted when people started saying, "I've been complaining about this for three years!" It wasn't high-tech. But it was human.

Once the team co-created the workflow, we layered AI on top. Scheduling. Invoice routing. Reporting. The stuff nobody wanted to do manually anyway.

Ninety days later? Productivity up 30%. Turnover dropped. Most importantly, people felt heard for the first time in years.

Where Human Meets AI

The story isn't about AI. It's about you.

The people using these tools are the ones creating the future. The magic happens when you combine human intuition with machine precision.

Leaders who see AI as a partner instead of a replacement? They're going to completely redefine what efficiency means.

"Automate the routine, humanize the exceptions." That's not a tagline. It's how you should be thinking about everything.

Every minute AI saves you is an opportunity. Deepen relationships. Serve customers better. Innovate faster. Do the work only humans can do.

What Leadership Looks Like Now

In this economy, leadership means curating, not managing.

The best leaders I know act like editors. They decide what to keep, what to revise, what to completely retire. They ask better questions.

"Which part of our DNA is helping us grow? Which part is turning into something that doesn't serve us anymore?"

AI helps you spot patterns faster. But acting on them requires courage. The courage to challenge legacy systems. Unspoken rules. Your own habits.

AI gives you insights. Leadership drives transformation.
Those are two different things.

Where We Go From Here

When your DNA changes, your worldview shifts.
When your business DNA changes, your entire ecosystem transforms.

This chapter is about recognizing that the story you've been telling might not be true anymore. And deciding what you're going to do about it.

That's the first step in recoding.
It starts right now.

CHAPTER 2

From Panic to Playbook

The Truth You Need to Hear First

Despite implementing AI. Despite cutting costs. Despite streamlining operations. Despite doing many of the things I'm about to teach you in this book, my plumbing company still failed.

We lost everything we'd built.
So why should you listen to me?

Because I learned these lessons the hard way, so you don't have to.

I waited too long. I ignored red flags for years. I kept pouring my speaking income into a sinking ship, telling myself it would get better. I convinced myself that one more quarter, one more hire, one more strategy would turn things around. By the time I started using AI strategically, the damage was already done. The debt was too deep. The hole was too wide. And my husband's health crisis forced us to face a truth we'd been avoiding for years: this business was killing us.

But here's what AI did give us: it gave us a dignified exit.

We weren't able to save the business, but we were able to sell it. The AI improvements made us operational, documented, and transferable. The new owners got functioning systems instead of chaos. Our employees kept their jobs. And William and I walked away with our reputations intact and a path forward.

Without AI, we would've just closed the doors on everything. Instead, we handed over a business that actually worked. It just didn't work for us, not anymore.

This book isn't about how AI saved my business. It's about how AI can save yours if you don't wait as long as I did.

When the Numbers Don't Lie

My husband, William, has always been relentlessly optimistic. He's the kind of person who finds meaning in near-death experiences and believes everything happens for a reason. That optimism, beautiful as it was, became dangerous when it came to his health.

He'd had asthma his whole life, but in his fifties, the lung infections started coming more frequently. Instead of seeing a pulmonologist, he'd do virtual doctor visits, get another round of antibiotics, and move on. I didn't realize how bad things had gotten until we started tracking his oxygen levels.

We connected his smartwatch to my phone so I could monitor his blood oxygen absorption in real time. Normal levels sit between 95% and 100%. Athletes might dip to 90% during intense exercise. William was running under 90% for five to six months and not telling anyone. Some days, his levels dropped below 80%.

Below 80% can kill you.

When I confronted him with the data, he shrugged it off. "I feel fine," he'd say, even though he was exhausted, foggy, and struggling to climb stairs without stopping halfway up.

I finally took control and made an appointment at National Jewish Health, one of the top respiratory hospitals in the world. It's right here in Denver. People fly in from across the globe to be treated there. They almost never admit patients for observation.

They admitted William for nine days.

Think about that. One of the world's premier respiratory hospitals looked at his condition and said, "This man needs to stay here." That's when I knew we'd been living in denial.

The Lightning Strike

When he was released, we made a pact: we'd work less, live more, and stop pretending the business wasn't draining us. We bought a camper and started taking weekends to explore Colorado, to breathe. Literally and figuratively.

One weekend, we camped at over 10,000 feet near Colorado Springs. On our way home, we pulled into a campground in a torrential rainstorm to dump the tanks. I stayed in the truck with the dogs while William went outside.

Then came the thunder.

It wasn't just loud. It shook the entire truck. The dogs yelped. My chest tightened. I looked out the rain-streaked window and saw William walking back toward me, soaked, pale, moving slower than I'd ever seen him move.

He tapped on the glass.

"I was just hit by lightning."

I stared at him through the window, rain hammering the windshield. "We're going to the hospital. Now."

He refused.

Of course he did.

He never went to the doctor. He never got checked out. But in the weeks that followed, his memory (already impaired from months of low oxygen) got worse. He was exhausted. Disoriented. His speech would slow mid-sentence, like his brain was buffering.

I knew it was the lightning. I have a traumatic brain injury myself from ten concussions and nineteen broken bones. That's a story for another book, but it means I recognize the symptoms when I see them. William wouldn't admit it, though. He just kept pushing through.

The Breaking Point

That's when we finally got honest with each other.

We sat in our living room one night, too tired to keep pretending. The plumbing company, the business we'd built together, wasn't just hard. It was crushing us. We were deeper in debt every month. I was using my speaking income just to cover payroll. For years, I'd felt like the company was stealing my energy, my ideas, and my money. And now it was stealing my husband's health.

We didn't enjoy it anymore.
We couldn't sustain it anymore.
We had to get out.

But we couldn't just walk away. We owed too much to too many people. We needed to make the company sellable, fast, without adding more stress or more hours to our already unsustainable schedule.

That's when I turned to AI.

Not because it was trendy. Not because I wanted to be an early adopter. Not because I believed some techno-utopian fantasy about robots saving the world.

Because we were desperate. And it was our last option before total collapse.

Facing the Truth About What We'd Built

When you're drowning, you don't see the water. You just know you can't breathe.

For years, I knew the plumbing company was broken, but I couldn't articulate how. I was too busy firefighting. Covering payroll gaps. Managing customer complaints. Training new hires who'd quit within weeks. Chasing invoices that should've been paid months ago.

But when William's health forced us to step back, we finally saw the machinery clearly. And what we saw was a mess.

I pulled out a whiteboard and did something I should've done years earlier: I mapped out where our time, money, and energy were actually going.

Here's what I found:

Time Drains

Bookkeeping: 12+ hours per week manually entering invoices, matching receipts, reconciling accounts. Our bookkeeper was talented but overwhelmed, doing work that didn't require a human brain.

Customer Service Chaos: Every phone call went to the same three people. Customers with billing questions got the same treatment as emergency service calls. Our team was exhausted, and our response times were embarrassing.

Onboarding Nightmare: New hires took 14+ hours of training, spread over two weeks. We had over 2,000 SOPs. Yes, two thousand. They were scattered across Google Drive, Dropbox, old email threads, and paper binders stuffed in filing cabinets. Nobody could find anything. New employees would ask the same questions over and over because we had no centralized system.

Marketing Inconsistency: We'd post on social media when we "had time," which was never. Our content was sporadic, generic, and forgettable. We had no brand voice, no content calendar, no system. We were invisible online, which meant we were losing leads to competitors

Energy Depletion

This was the hardest one to quantify, but it was the most destructive.

William was working 60+ hour weeks while his oxygen levels plummeted.

I was spending my speaker income (money I'd earned on stages across the country teaching other people how to run their businesses) to cover payroll gaps at the plumbing company. I was miserable and everyone knew it.

Our team was burned out and unmotivated. Turnover was high. Morale was low. The good people left. The mediocre ones stayed because they couldn't get hired anywhere else.

We were too exhausted to innovate, strategize, or even enjoy the rare moments when things went well.

The brutal truth: We had built a business that required us to function, but it was killing us in the process.

And we'd done it for too long. The debt had compounded. The inefficiencies had metastasized. We'd been running a dying company on life support, pretending the monitor wasn't beeping.

The Question That Changed Everything

I stared at that whiteboard for a long time. Then I asked myself a question I'd been avoiding for years:

"If we were starting from scratch today, would we build it this way?"

The answer was immediate and obvious: Hell no.

We wouldn't manually enter invoices. We wouldn't make every new hire read through 2,000 disorganized documents. We wouldn't let phone calls randomly interrupt whoever happened to be nearby. We wouldn't post on social media only when we felt inspired.

But we were doing all of those things. Not because they worked, but because that's how we'd always done them.

Our business DNA was outdated. And it was making us sick.

But here's the thing: even as I saw all this clearly, I knew the truth underneath the truth.

We were too far gone.

AI could make us efficient. It couldn't erase years of accumulated debt. It couldn't undo poor financial decisions. It couldn't repair a business model that was fundamentally unsustainable in our market.

What it could do was give us an exit with dignity.

What AI Actually Did for Us

Over the next eighteen months, we used AI to:

- Automate bookkeeping workflows so invoices, matching, and reconciliation happened in minutes instead of hours.
- Systematize onboarding so new hires could learn from our 2,000 SOPs without drowning in them. (Using Trainual)
- Create consistent marketing so we had a content calendar, brand voice, and visible presence online.
- Route customer calls intelligently so the right person handled the right inquiry without overwhelming anyone. (outsourced)

We didn't save the business. But we made it functional.

We documented everything. We created systems that could run without us micromanaging every detail. We reduced operational waste significantly, enough that when buyers looked at the books, they saw a business worth purchasing, not a liability to avoid.

And here's what matters: The employees that wanted to, kept their jobs and our customers had someone to call.

William now works part-time for them, three days a week, on his terms, with his oxygen levels stable and his stress manageable. Our team didn't get laid off in a dramatic collapse. They got new leadership and a fresh start.

We walked away at the end of our rope, yes. But we walked away with our heads up, not in shame.

Coming Home: From AT&T to Plumbing and Back Again

Here's what I haven't told you yet: the plumbing company was never supposed to be my career.

Before William and I bought that business, I was an international General Manager with AT&T Wireless, responsible for network operations in two Caribbean countries. I was part of the team that rolled out wireless networks when most people still had landlines. I carried one of the first camera phones, and trust me, people thought it was ridiculous. "Why would anyone need a camera on their phone?"

I was on the wireless data team when everyone said mobile internet was a scam. "No one will want to check email on a tiny screen." "Data plans will never work." "It's too expensive, too slow, too complicated."

Sound familiar?

It's the exact same conversation happening around AI today.

I've seen this movie before. I know what technology adoption looks like when it's early, messy, and polarizing. I know what separates the people who adapt from the people who get left behind. And I know that the companies that win aren't the ones with the best technology. They're the ones with the best implementation strategy.

Why I Left and Why I Came Back

So how did I go from international telecom executive to plumbing company owner?

Life happened. Family needs. A belief that we could build something together. And honestly? Burnout. Corporate life had chewed me up, and I wanted to build something that felt more tangible, more human, more ours.

For a while, it worked. But somewhere along the way, we stopped building and started just surviving.

When the business started failing, I had a choice: keep pretending I was a plumber, or go back to what I actually know. Technology, systems, transformation, and how to help people navigate change without losing their minds.

So I went back to speaking. Back to consulting. Back to the work I was doing before the plumbing company consumed my life.

Except now, I had something those early years at AT&T never gave me: I understood what it's like to be the person resisting change.

At AT&T, I was the one rolling out the technology. I was the insider, the evangelist, the "of course this will work" person. I didn't fully understand the fear, the skepticism, the exhaustion of people on the ground trying to make new systems work while still doing their jobs.

But running the plumbing company? I became the person I used to try to convince. The one who's tired, overwhelmed, and thinking, "I don't have time for another new thing."

That's the experience that makes me qualified to write this book.

Not because I succeeded spectacularly. Because I failed, and learned, and came back with a framework that actually works for real businesses run by real, exhausted people.

The Speaking Circuit Saved Us (Literally)

While the plumbing company was drowning, my speaking business kept us afloat. Barely.

I was traveling to conferences, corporate events, and leadership retreats, teaching people about digital transformation, AI adoption, and change management. The irony wasn't lost on me: I was teaching other companies how to use technology strategically while my own business was hemorrhaging money because we weren't.

But here's what the speaking work gave me: access to hundreds of companies trying to figure out the exact same thing.

I got to see what worked and what didn't across industries, company sizes, and leadership styles. I saw the mistakes in real time. I watched teams resist, adopt, thrive, or implode. I tested strategies with clients that I wished I'd implemented years earlier in my own company.

And slowly, I built a framework. Not from theory, but from pattern recognition across dozens of struggling businesses.

That framework became the backbone of how we automated the plumbing company enough to sell it. And now, it's the backbone of this book.

The Parallel No One Talks About

When I was at AT&T rolling out wireless data, people said:

"It's too expensive."
"It's not secure."
"No one needs it."
"It'll never replace what we already have."
Today, with AI, people say:
"It's too expensive."
"It's not secure."
"No one needs it."
"It'll never replace what we already have."

The resistance isn't about the technology. It's about fear of obsolescence.

And the companies that won the wireless revolution weren't the ones with the best phones or the fastest networks. They were the ones who made adoption easy. Who met people where they were. Who built trust through transparency and small wins.

That's the lesson I brought from AT&T to the plumbing company, just way too late.

And that's the lesson I'm bringing to you now: AI adoption is a people problem, not a technology problem.

The tools work. The question is: will your people use them? Will they trust them? Will they see AI as a teammate or a threat?

That's where I come in.

I've been the executive mandating change. I've been the business owner resisting it. And now, I'm the consultant helping companies bridge that gap.

I've lived all three sides of this transformation. And that's why I know what I'm talking about, even though the business broke me before I could fix it.

Actually, especially because of that.

Why I'm Telling You This

I could have written a book filled with glossy success stories and seven-figure exits. I could have pretended AI saved my business and turned me into a millionaire.

But that would be a lie.

And lying would make me just another guru selling false hope.

The truth is harder and more valuable.

AI won't save a fundamentally broken business. It won't erase bad debt, fix broken business models, or rescue you from decades of poor decisions.

But AI will give you leverage. It will buy you time. It will help you see clearly what's working and what isn't. And if you're in a position like I was (trying to exit with dignity), it will make you sellable instead of just closable.

More importantly, if you start earlier than I did, AI can prevent you from ever getting to where I was.

If I'd implemented these strategies three years earlier, when the business was still viable, before the debt compounded, before William's health collapsed, we might still own that company. We might have turned it around.

But I waited. I told myself it would get better. I kept pouring good money after bad. I avoided the hard truth until the truth became unavoidable.

Don't be me.

Use this book to:

- Identify operational waste before it buries you
- Automate the repetitive work that's stealing your energy
- Create systems that work without you being the bottleneck
- Make better decisions faster by seeing your data clearly
- Build a business that serves your life, not consumes it

I can't give you a fairy tale. But I can give you a roadmap drawn in scar tissue and hard-won wisdom.

And sometimes, that's worth more.

The Turning Point

AI became that plan.

We began using AI not as a shiny new tool, but as a lifeline. It helped us reimagine processes, reduce waste, and find efficiencies we didn't think were possible. We cut our operating costs exponentially without cutting people, and rebuilt the business in a way that made it sellable. It wasn't the graceful exit we had dreamed of. In truth, we barely got out alive. But in hindsight, it was the best thing that ever happened to us.

My husband now works part-time for the new owners, three days a week, as his health allows. Our relationship is stronger than ever, our lives calmer, and our purpose clearer. That journey made me an evangelist for AI. Not just the technology, but the freedom it creates when you use it right.

Digital Denialists: A New Breed of Resistance

But not everyone is ready for this kind of transformation. In fact, I've given a name to a growing group of people I see in almost every audience and organization: Digital Denialists.

These are the individuals who refuse to get on board with their company's AI or technology strategy. Some are intimidated and fear they'll lose their jobs. Others simply don't want to learn. But their resistance isn't harmless. It's contagious. Digital denialists slow innovation, stall progress, and, in some cases, actively damage their organizations' ability to compete.

In the era of AI, refusing to adapt is no longer neutral. It's a liability. While some teams are using AI ecosystems to cut costs, improve workflows, and boost productivity, the denialists cling to the old ways, insisting that "this too shall pass." It won't. The companies that thrive will be those that bring their people along for the journey, creating a culture that sees AI not as a threat, but as a teammate.

The Five Types of Digital Denialists

Over hundreds of conversations with leaders and teams, I've identified five distinct types of digital denialists. Recognizing which type you're dealing with is the first step to addressing their resistance effectively.

1. The Skeptic: "I'll Believe It When I See It"

Profile: Usually a high performer who's been burned by past "flavor of the month" initiatives. They've seen CRM rollouts fail, project management tools gather dust, and training programs fizzle out. Now they're applying that same lens to AI.

What they say:

"We tried automation before. It didn't work."
"This is just another shiny object."
"I'll wait until it's proven."

What they're really thinking: If I invest energy in this and it fails, I'll look foolish.

How to spot them: They attend meetings but contribute minimally. They ask pointed questions designed to expose flaws. They're polite but unmoved.

The conversion strategy: Show, don't tell. Skeptics need proof, not promises.

Give them a micro-win task: "Try this one prompt for your weekly report and tell me if it saves you time."

Share data from peers, not leadership: "Sarah in accounting cut her reconciliation time by 40%."

Invite them to shadow a pilot: Let them see it working in real conditions.

Success story: I worked with a finance director who rolled her eyes at every AI mention. I asked her to try one prompt to summarize her monthly variance report. She begrudgingly agreed. The AI did in 3 minutes what took her 45 minutes. The next week, she brought me five more processes to automate. Skeptics become your best advocates once they're converted because they did the math themselves.

2. The Perfectionist: "It's Not Ready Yet"

Profile: Detail-oriented, conscientious, often in quality control or compliance roles. They see every edge case, every potential error, every reason to wait.

What they say:

"The technology isn't mature enough."
"What if it makes a mistake?"
"We need more testing before we roll this out."

What they're really thinking: If something goes wrong on my watch, I'm responsible.

How to spot them: They volunteer to be on committees but never vote to move forward. They request "just one more pilot" indefinitely. They focus on theoretical risks over actual results.

The conversion strategy: Reframe perfection as iteration.

Use the phrase: "We're running experiments, not launching final products."

Give them control of the guardrails: "You design the review process. You decide what needs human oversight."

Show them how AI reduces errors in repetitive tasks: "AI doesn't get tired at 4pm on a Friday."

Success story: A compliance officer in a healthcare company resisted AI transcription for patient notes, citing accuracy concerns. We ran it in shadow mode for 60 days. AI transcribed, humans reviewed. The AI error rate was 2.3%. The human error rate? 4.1%. Data convinced her. Now she's the internal champion for AI documentation.

3. The Ghost: "I'll Get to It Later"

Profile: The person who nods in meetings, says they'll try the new tool, then never logs in. They're not openly resistant. They just quietly opt out.

What they say:

"I've been so busy, I haven't had time."

"I'll look at it next week."
"Can you send me that link again?"

What they're really thinking: If I ignore this long enough, it will go away.

How to spot them: Zero activity logs. No questions asked. They're present but invisible.

The conversion strategy: Make it unavoidable, but easy.

Embed AI into their existing workflow: Don't add a new tool. Integrate it into the tool they already use daily.

Assign a buddy: Pair them with an early adopter who can walk them through it.

Set a small, public commitment: "By Friday's team meeting, share one thing you tried."

Success story: A sales manager ghosted our CRM AI assistant for three months. We finally integrated it directly into his email. He didn't have to go anywhere new. The AI started drafting follow-up emails based on his meeting notes. Within a week, he was using it for every client interaction. Sometimes the barrier isn't resistance. It's friction.

4. The Martyr: "I Guess I'm Just Old School"

Profile: Often a long-tenured employee with deep institutional knowledge. They wear their resistance as a badge of honor, positioning themselves as the keeper of "how we really do things."

What they say:

"I've done it this way for 20 years."
"The personal touch matters more than speed."
"You can't automate relationships."

What they're really thinking: My value is in my experience. If AI can do what I do, what's left?

How to spot them: They frame their resistance as moral superiority. They contrast "real work" with "letting a robot do it."

The conversion strategy: Elevate their expertise, don't replace it.

Position AI as amplifying their wisdom: "Your 20 years of knowledge could train an AI to help the whole team."

Give them a teaching role: "We need you to review AI outputs and flag what it's missing."

Highlight what only they can do: "AI handles the routine. You handle the exceptions, where judgment matters."

Success story: A customer service veteran resisted AI chatbots, insisting customers deserved "real people." We asked her to train the chatbot on her best responses and review its interactions weekly. She discovered she loved teaching the AI and correcting its tone. She now sees herself as a coach, not a competitor. Her line: "The bot handles the easy stuff so I can focus on the customers who really need me."

5. The Saboteur: "This Will Never Work Here"

Profile: The most dangerous type. Actively undermines AI initiatives through cynicism, misinformation, or passive-aggressive compliance.

What they say:

"Leadership doesn't understand our work."
"They're just trying to cut jobs."
"I heard this failed at [other company]."

What they're really thinking: If this succeeds, it proves I was wrong. I can't let that happen.

How to spot them: They poison the well in hallway conversations. They spread fear without offering solutions. They comply minimally while vocally predicting failure.

The conversion strategy: Address directly, but privately.

Have the hard conversation: "I've noticed you're expressing concerns. Let's talk about what's really going on."

Offer an exit ramp: "If you're not bought in, I need to know now. What would it take for you to support this?"

Set clear expectations: "You don't have to be a champion, but I do need you to stop undermining the team."

Sometimes, separation is necessary: If they won't engage or adapt, they may not belong in your future organization.

Success story: A department manager openly mocked AI initiatives in team meetings, calling them "Silicon Valley nonsense." After two private conversations with no shift, leadership gave him a choice: lead his team through adoption or transition to a different role. He chose to leave. Six months later, his replacement had the department running 30% more efficiently. The rest of the team admitted they'd felt held hostage by his negativity. Sometimes, removing one blocker unlocks everyone else.

The Pattern Behind the Resistance

Regardless of type, digital denialists share one common trait: fear of irrelevance.

They're asking: If AI can do what I do, who am I?

Your job as a leader isn't to dismiss that fear. It's to reframe the answer.

The truth is: AI can't do what you do. It can do what a process does. It can handle the repetitive, the routine, the predictable.

But it can't:

- Navigate a customer crisis with empathy
- Read the room in a negotiation
- Mentor a struggling team member
- Spot an opportunity no one else sees
- Make a judgment call when the rules don't apply
- AI doesn't replace people. It replaces tasks.

And when you help your team see that distinction, resistance transforms into curiosity.

Your Action Plan

This week, identify the denialists in your organization:

- List their names and types (Skeptic, Perfectionist, Ghost, Martyr, Saboteur)
- Assign each a conversion strategy from above
- Start with the easiest wins: Skeptics and Perfectionists respond to data
- Address Saboteurs quickly: Don't let them spread fear
- Celebrate every conversion publicly: "Here's how [Name] is using AI now..."

Remember: You don't need 100% adoption on Day 1. You need enough momentum that the holdouts feel left behind, not left out.

What This Means for You

If you're reading these words and thinking, "That sounds like my business," you're not alone.

Duct tape and heroic effort hold most companies together. Most leaders work within the business, not on it. Most teams are drowning in tasks that could (and should) be automated.

But here's the positive news: you don't have to redesign everything at once.

You just need to start with one workflow. You only need to focus on one process that is repetitive, soul-crushing, and time-consuming. Automate that. Measure the results. Then move to the next one.

That's the playbook. And in the chapters ahead, I'm going to show you exactly how we did it. Step by step, tool by tool, mistake by mistake.

Because if we could dig ourselves out of debt, save $170,000 a year, and sell a dying company to new owners who actually wanted it, you can transform your business too.

Let's get to work.

Your Turn

Before we go further, I want you to ask yourself the same question I had to answer in that living room:

What is your business costing you that you can't afford to pay anymore?
Is it your health? Your relationships? Your joy? Your future?

And here's the harder question:

Are you already too far gone, or do you still have time to turn it around?

If you're reading this and recognizing yourself in my story, don't wait. Don't tell yourself it'll become better on its own. Don't keep pouring good money after bad. Don't sacrifice your health and your family for a business that's already dying.

Be honest. Get help. And if AI can provide you leverage to turn things around (or at least exit with dignity), then use it.

The worst thing I did wasn't losing the business.

The worst thing I did was waiting so long before facing the truth.

Exercise: Your Wake-Up Call Assessment

Take 10 minutes right now and answer these questions honestly:

Health Check: Is your business affecting your physical or mental health? Your family's?

Financial Reality: Are you using personal funds to cover business expenses? How long has this been happening?

Energy Audit: Do you feel energized by your business, or drained by it?

The Honest Question: If you could start over knowing what you know now, would you build this business again?

The Timeline: If nothing changes, how long can you sustain this? Six months? A year? Three years?

If your answers scare you, good. Fear is data. Use it.

Signature Elements

Tour Guide Tip: The best time to fix a business is before it's broken. The second-best time is now.

Guardrail: Don't confuse efficiency with viability. AI can make a bad business run smoothly, but it's still a bad business.

Try It in 10 Minutes: List three things your business does that drain money, time, or energy. Be brutally honest. That's your starting point.

From Old School to New School: The Foundation

PART II

The Recode Process

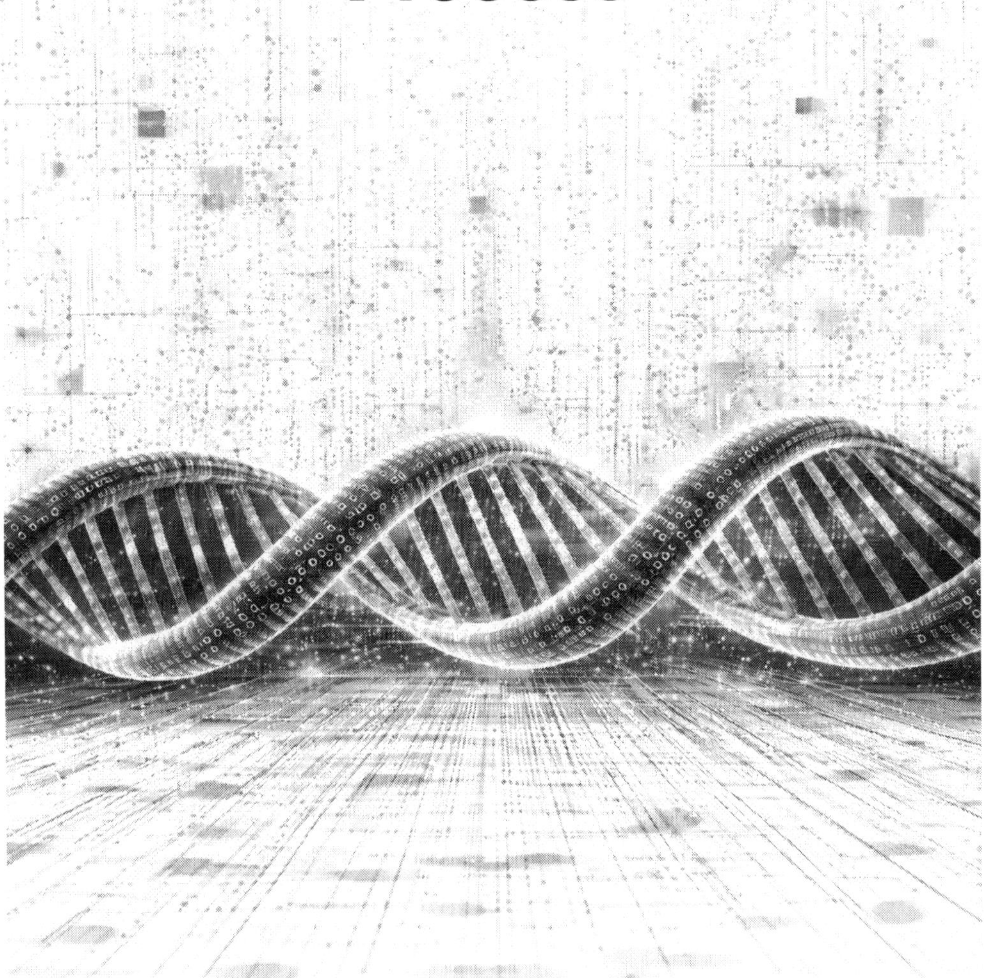

CHAPTER 3

The Three Processes (Start with This)

Start where it matters most. Not with some flashy AI campaign. Not with a custom avatar that looks vaguely like you but with dead eyes. Start where it actually counts.

I tell every audience the same thing:

"You don't start with AI in your marketing. You start with AI in your money, your operations, and your momentum. Because if your house isn't in order, AI just multiplies the mess."

I learned this the hard way. Spent three months building an elaborate AI content system before fixing our backend processes. Know what happened? We created beautiful marketing for a business that couldn't fulfill orders efficiently. Brilliant.

These are the three workflows I'm about to walk you through. The foundation every business owner and C-suite leader should automate first. Each one frees time, cuts waste, and gives your team energy back for innovation and actual service. Get these right? Everything else gets easier. Ignore them? Good luck.

But here's what nobody tells you at those $2,000 AI conferences: the technology? That's honestly the easy part.

Leadership is where everything falls apart or comes together.

You can drop $50,000 on the best AI tools in the world. Top-tier stuff. Doesn't matter one bit. If your people don't trust the process, don't understand the why, or think they're about to be replaced, your transformation will crash and burn before it starts. I've watched it happen. Multiple times. In companies with way bigger budgets and supposedly smarter people than mine.

AI implementation isn't a tech problem. Never has been. It's a people problem. Always has been. And people problems require actual leadership, not software licenses and Slack announcements.

This chapter gives you both. The workflows and the leadership framework to make them stick.

The Leadership Imperative: Why AI Adoption Lives or Dies at the Top

Let me be blunt. If you're a leader who thinks you can delegate AI transformation to IT or "the tech person," you've already lost.

I've watched this disaster play out at least two dozen times in the last year alone. CEO announces an AI initiative at the quarterly all-hands. Big vision. Lots of energy in the room. Maybe even brings in donuts. Then hands the whole thing off to the CTO or the director of operations who already has 47 other priorities. Six months later? Nothing's changed except everyone's more frustrated, the CTO is secretly updating their LinkedIn, and the CEO is standing in the break room wondering out loud why "nobody wants to innovate around here."

Here's why that happens every single time: AI adoption isn't some side project you can toss over the wall to someone else and check off your list. It's a complete business transformation that touches every department, every process, and every single person who shows up to work. And transformation? That doesn't happen from the middle. Never has. It happens from the top.

Your team is watching you right now. Trust me on this. They're watching everything. They're watching to see if you're actually serious or if this is another passing fad. They're watching to see if you're using AI yourself or if you're just mandating it for everyone else while you keep doing things the old comfortable way. They're watching to see if this is real change or just another flavor-of-the-month initiative that'll blow over by Q3 when something shinier comes along.

Here's what I learned from scaling three different businesses and leading AI rollouts across companies ranging from 15 to 200 employees: your adoption rate will never, ever exceed your leadership commitment.

If you're tentative about it, they'll be resistant.
If you're inconsistent, they'll straight-up ignore it.
If you're absent, they'll sabotage it. Sometimes on purpose. Usually by accident. The result's exactly the same either way.

But if you actually lead it, learn it alongside them, and model it in your own work? Your team will follow. Not because they have to. Because you showed them it was worth the effort.

The Five Leadership Principles for AI Adoption

1. Lead from the Front, Not from the Memo

Don't announce an AI initiative in some corporate all-hands email and expect your team to just run with it. That's not leadership. That's abdication with a professional subject line.

Show them. Use AI in your own work first. Share your wins. Share your embarrassing failures. Let them watch you learn in real time, mistakes and all.

When I rolled out AI in our service business back in early 2023, I didn't start with some big companywide mandate and training sessions. Started with myself. Just me. Used AI to automate my own weekly reports that I genuinely hated writing. My meeting prep that ate up three hours every Monday. That's leadership. Not delegation. Participation.

And yeah, I messed up constantly. Sent a few AI-generated emails that were so off-brand I had to send immediate apologies. Had to completely redo a financial report because the AI hallucinated a statistic about our HVAC revenue that sounded plausible but was totally made up. My team saw every single failure. The wins and the spectacular face-plants. And you know what that actually did? Made them way more willing to try it themselves. Because I wasn't standing there pretending to be perfect or acting like I had it all figured out from day one.

2. Communicate the Why, Not Just the What

People resist change when they don't understand the purpose behind it. They need to know why this matters to them personally, not just what they're supposed to start doing differently this Monday morning.

Don't walk in and say: "We're implementing AI in customer service."
Try saying: "We're implementing AI so you can stop answering the exact same question 47 times a day about whether we service water heaters in Littleton and start solving the complicated problems that actually require your expertise and 10 years of experience."

See the difference? One sounds like a mandate from corporate that's going to make your life harder. The other sounds like freedom.

Frame AI as a tool that gives them their time back, their sanity back, their ability to do the work they were actually hired to do and presumably care about. Not a tool that monitors their productivity every second or threatens their job security. Because if they think it's the latter? They'll resist it with absolutely everything they've got. And they'll win. I've watched entire teams tank AI initiatives through nothing but passive resistance. Never underestimate the power of people who genuinely don't want to change.

3. Build Trust Through Transparency

Your team needs to know what AI is actually doing behind the scenes, how it's being used day-to-day, and what decisions are still completely theirs to make. Opacity breeds fear and conspiracy theories. Transparency builds trust. This isn't complicated, but somehow leaders keep getting it spectacularly wrong.

Be crystal clear about:
What specific tasks AI will handle
What tasks absolutely still require human judgment
How performance will actually be measured
What success looks like in concrete terms

When we automated invoice processing in late 2022, I didn't just flip a switch one Monday morning and announce it was done. Spent two full weeks walking the accounting team through every single step of the process. Showed them exactly what the AI was doing, what it was flagging for review, where they still had complete control. Let them test it extensively. Question everything. Poke holes in it. Refine all the parameters based on their feedback.

They went from deeply skeptical to genuinely enthusiastic in about three weeks. Not because the technology was some kind of magic (though it was pretty impressive). Because they understood it, trusted it, and could see with their own eyes it was making their lives easier without making them expendable or obsolete.

4. Create Psychological Safety for Experimentation

AI adoption requires experimentation. Real experimentation. Experimentation requires failure. Lots of it. And failure requires an environment where people aren't terrified to try something new because they might get it wrong and end up in a performance review or worse.

Make it genuinely safe to test new approaches. Make it safe to fail without consequences. Make it safe to walk into a team meeting and say out loud, "I tried this prompt everyone's talking about and it completely bombed. Total disaster. Here's what I learned though."

I started telling my teams something that sounded crazy at first: "If you're not breaking at least one AI tool a month, you're not experimenting hard enough." Sounds ridiculous. But it worked like magic. People started taking real risks. Trying genuinely new approaches. Finding solutions I would've never thought of in a million years because I was stuck in my old ways of thinking.

That explicit permission to fail? That's what turns hesitation and fear into actual innovation. Without it, you get compliance but you never, ever get creativity or breakthroughs.

5. Celebrate Early Wins and Share Them Widely

Nothing builds momentum faster than proof that actually works in your specific environment with your actual people using your systems. When someone on your team uses AI to save three real hours on a real project, shout it from the rooftops. Share the exact prompt they used. Share the whole process step-by-step. Share the before-and-after screenshots. Make them the absolute hero of the week.

Early wins create believers. Believers create passionate advocates. Advocates create organic culture change that spreads naturally instead of through top-down mandates that everyone resents.

We had this customer service rep named Maria back in summer 2023. Started using AI to draft responses to our ten most common customer questions. Cut her response time literally in half. We celebrated that specific win in our next company meeting. Showed everyone her prompts and her results. Gave her a $100 gift card to her favorite restaurant. Next week? Five more people were experimenting with AI in their own workflows without anyone telling them to. Week after that? Twelve people. We didn't mandate it. We celebrated it genuinely, and it spread like wildfire.

That's how real adoption actually happens in the real world. Not through mandates and policies. Through proof that it works and recognition for the people who make it work.

The Leadership Playbook: Rolling Out AI Systemically

Here's the actual step-by-step process I use with every single organization I work with now. This isn't theory I read in some Harvard Business Review article written by consultants who've never run a business. This is what actually works when you're implementing AI with real people who have real jobs and very real skepticism.

Step 1: Start with a Pilot Team, Not a Companywide Rollout

Choose 3-5 people who are naturally curious, reasonably tech-comfortable, and genuinely influential within your organization. Not necessarily your top performers. Not necessarily your managers. Your connectors. The people other people actually listen to in the break room and trust. These become your AI champions. Train them intensively first. Let them experiment freely. Let them find both the wins and the landmines before anyone else.

When they succeed, the rest of your team will want in. Social proof from peers is infinitely more powerful than any mandate from leadership.

I learned this lesson the expensive way. In my first AI rollout, we brought in catered lunch. Tried to teach 50 people how to use basic AI tools in three hours. Total unmitigated disaster.

People got completely overwhelmed. Questions went unanswered because I physically couldn't keep up. Half the team gave up before they even got started because it felt too complicated.

Second rollout, completely different company? Pilot team of five carefully chosen people first. Trained them intensively over two weeks. Let them become genuine experts. Then they trained everyone else in smaller groups. Worked like absolute magic.

Step 2: Identify High-Impact, Low-Risk Use Cases

Don't start with your most critical process. Don't start with anything customer-facing. That's asking for trouble. Start with something repetitive, easily measurable, and genuinely low-stakes. Something where failure won't break your business but success will be obvious to everyone watching.

Good first projects to try:

- Automating meeting summaries and action items
- Generating email responses to common internal questions
- Creating first drafts of internal reports or proposals
- Summarizing customer feedback from surveys or reviews

Bad first projects to avoid:

- Replacing your entire CRM system in one shot
- Automating client-facing communications without thorough human review
- Anything involving sensitive data before you have rock-solid governance in place
- Anything your CEO personally touches or cares deeply about

Step 3: Document and Systemize What Works

When you finally find a process that actually works reliably, don't let it live in one person's head or buried somewhere in their personal ChatGPT history where nobody else can access it. Document it carefully. Turn it into a genuinely repeatable system that literally anyone can follow step-by-step.

Create these things:

- Prompt libraries (organized logically by department and specific use case, with working examples)
- Process maps (detailed step-by-step guides for common AI workflows)
- Review gates (crystal clear about where human oversight is absolutely required)
- Success metrics (how you'll actually measure real impact, not just activity)
- This is how you go from one person using AI to save themselves time to your entire organization using it consistently and effectively at scale.

And yeah, this part takes real time. It's genuinely tedious work. Nobody wants to do it. But you know what's infinitely more tedious? Training the exact same process 47 different times to 47 different people because nobody bothered to write down how it actually works.

Step 4: Train in Waves, Not All at Once

Trying to train 200 people on AI in one massive all-company session is a disaster waiting to happen. I've done it. Twice. Do not recommend. You'll overwhelm everyone, confuse them completely, and create way more resistance than adoption.

Instead, train in deliberate waves:

- Wave 1: Your pilot team (the champions who'll eventually become your internal trainers)
- Wave 2: Department leads and managers (who need to support their teams effectively)
- Wave 3: Frontline teams, with training tailored specifically by department and actual role

Each wave learns directly from the one before. By the time you finally get to Wave 3, you're teaching proven processes with real examples from your own company, not theoretical possibilities from someone else's case study that might not even apply to you.

Step 5: Build Feedback Loops and Iterate Constantly

AI adoption isn't a one-and-done project you can check off your Q2 goals and forget about. It's an ongoing process that needs constant attention and adjustment. You need real feedback loops to know what's actually working in practice, what's broken, and where people are genuinely stuck or quietly giving up.

Set up:

- Weekly check-ins with your pilot team (15 minutes max, standing meeting)
- Monthly town halls where people share both wins and challenges openly (make it genuinely safe to share failures)
- Anonymous feedback surveys to surface concerns people absolutely won't voice publicly (people will tell you things anonymously they'd never say in a meeting)
- Office hours where literally anyone can drop in and get help with their AI experiments (I did mine every Thursday at 2pm for six months straight)
- Listen to the feedback. Actually listen, don't just collect it. Adjust your process based on what you hear. Keep iterating based on reality.

I can't tell you how many times someone raised a legitimate concern in an anonymous survey that completely changed how we rolled out a particular tool. Things I would've never thought of in a thousand years. Things that would've absolutely killed adoption company-wide if we hadn't caught them early and addressed them properly.

Step 6: Tie AI Adoption to Business Outcomes, Not Just Activity

Don't measure success by how many people are using AI or how many prompts they've run this month. That's pure vanity metrics that don't mean anything. Measure success by what AI is helping you actually achieve in your business that moves the needle.

Track these things:

- Time saved per process (in actual hours, not estimates or guesses)
- Cost reduction per department (in real dollars that show up on your P&L)
- Revenue generated through AI-assisted work (attributable and genuinely measurable)
- Employee satisfaction and retention (survey your team regularly and honestly)
- Customer satisfaction scores (did service actually improve or secretly decline?)

When you tie AI directly to real business outcomes that show up in your financials, it stops being a "nice to have" and becomes a "can't live without." That's when you finally get real buy-in from the people who control the budget and make actual decisions.

The Systemization Framework: From Chaos to Clarity

Systemizing your business with AI isn't about adding more tools to an already bloated tech stack nobody fully understands. It's about creating a solid foundation where processes are genuinely repeatable, actually scalable, and resilient enough to survive when your best person quits suddenly or goes on maternity leave.

Here's how to think about this:

1. Map Your Current State

You can't systemize what you don't understand. Start by mapping your existing workflows as they actually work today, not how you think they work or how the manual says they should work.

For each major process, honestly ask:

- Who actually owns this day-to-day when things go wrong?
- What are all the real steps, including the informal workarounds everyone uses?

- Where are the bottlenecks that slow everything down and frustrate people?
- What's manual that could potentially be automated without breaking things?
- What's inconsistent that desperately needs to be standardized?

Use AI to help with this mapping part. Feed it your existing SOPs, interview transcripts with your team, or screen recordings of processes being done in real time. Ask it to identify inefficiencies and bottlenecks. You'll be genuinely surprised what it catches that you've been completely blind to for years because "that's just how we've always done it."

PROMPT EXAMPLE: "Analyze this workflow and identify specific steps that are repetitive, time-consuming, or prone to human error. Suggest which steps could realistically be automated with AI and which absolutely require human judgment."

2. Prioritize Based on Impact and Ease

Not every process is worth automating. Some things genuinely should stay manual. Focus ruthlessly on the ones that give you the biggest return for the least effort and risk.

Create a simple 2x2 matrix:

- High Impact, Low Effort → Do these first (your quick wins)
- High Impact, High Effort → Plan these carefully for later phases
- Low Impact, Low Effort → Do these only if you have extra bandwidth
- Low Impact, High Effort → Skip these entirely, seriously not worth it

I wasted about four solid months in early 2023 automating processes that ultimately didn't matter at all because I didn't prioritize properly upfront. Learned that lesson the expensive way. Don't be me.

3. Build Systems, Not Shortcuts

The difference between a shortcut and a system is repeatability. Can someone else do it without asking you questions? Can it survive turnover? Can it scale beyond one person?

A shortcut is: "I use ChatGPT to draft my emails when I remember to."
A system is: "We have a library of 30 tested email prompts organized by use case, with clear review gates and quality standards, and everyone on the customer service team knows exactly how to use them and when."

Shortcuts help individuals save some time. Systems transform entire organizations permanently.

4. Embed AI into Your Standard Operating Procedures

AI shouldn't be separate from your processes like some optional add-on feature people can choose to use or ignore. It should be woven directly into how work gets done by default.

Update your SOPs to include:

- When to use AI (and when absolutely not to)
- Which specific tools to use for which exact tasks
- What prompts to use (with working examples people can copy)
- Where human review is absolutely required
- How to escalate issues or exceptions that come up

Make AI the default way of working, not something people have to remember to do separately.

5. Create Governance and Guardrails

Systemization without governance is just chaos with fancier technology. You need clear, written rules about what AI can and can't do in your organization.

Establish:

- Data privacy policies (what data can AI access? What's absolutely off-limits?)

- Quality standards (what level of accuracy is required before something goes live?)
- Approval workflows (who reviews AI-generated work before it's final?)
- Ethical guidelines (how do we use AI responsibly and transparently?)

This isn't about limiting AI or slowing people down unnecessarily. It's about using it with genuine confidence and control so you don't wake up one morning to a PR crisis because someone fed customer data into a public AI tool.

Workflow One: Finance

Cut Waste Without Cutting People

When AI comes to finance, it's not about replacing people. It's about replacing repetition. Your team's time is way too valuable to waste on manual matching, double-entry bookkeeping, and endless reconciliations that a computer could do in literal seconds. These tasks drain hours every single week that could be spent on strategy, actual analysis, and growing the business.

But here's the leadership piece nobody talks about at those fancy AI conferences where everyone's drinking overpriced coffee: your finance team is absolutely terrified right now.

Terrified that automation means their job is on the chopping block. Terrified of trusting a machine with numbers when their name is on the reports. Terrified of being personally blamed if the AI makes a mistake and screws up the books. I've had accountants cry in my office over this.

Your job as a leader is to address that very real fear head-on, not pretend it doesn't exist or dismiss it as irrational.

How to Lead Finance Automation:

1. Start the conversation with empathy, not efficiency.

Don't walk into your accounting department on a random Tuesday morning and announce, "Great news everyone, we're automating invoice processing to save time and cut costs!" That sounds exactly like, "We don't need as many of you anymore, start polishing your resumes and clearing out your desks."

Instead, try something like: "I know you're spending somewhere between 10-15 hours every week on data entry and reconciliation. That's genuinely not why I hired you. I hired you because you understand our financial health deeply and can help me make way better business decisions. Let's use AI to handle the repetitive data entry work so you actually have time to do the strategic analysis that moves the needle."

Big difference. Same outcome. Completely different emotional response from your team.

2. Involve them in the process, not after the decision.

Let your finance team actually test the tools themselves. Let them choose what to automate first based on what's driving them crazy. Let them set the quality standards and approval workflows based on their expertise. Give them real input, not the fake illusion of input.

When people have genuine ownership over the change, they support it enthusiastically. When change is forced on them from above, they resist it. Sometimes actively. Sometimes passive-aggressively. The result's exactly the same either way.

3. Make AI the junior analyst, not the replacement.

Position AI as the entry-level team member who does the grunt work while your experienced human team does the high-value strategic work that requires judgment and deep context.

AI handles: data entry, categorization, basic reconciliation, flagging anomalies for human review.

Humans handle: strategic analysis, actual decision-making, stakeholder communication, complex problem-solving, anything involving judgment calls.

That's a genuine partnership. Not a replacement. Frame it exactly that way from day one.

Playbook: Finance Workflow

1. Automate Invoice Management

Use tools like Dext, Ramp, or Bill.com to read and categorize invoices automatically. Zero human data entry required after the initial setup.

Leadership move: Have your finance lead personally test three different tools and recommend one based on your specific needs and existing systems. Their buy-in matters infinitely more than your personal preference.

> **IDEA ZONE**
>
> Every single time someone types, clicks, or copies the exact same thing twice, AI can probably do it once, faster and way more accurately. Start tracking repetitive financial tasks for just one week and highlight which could realistically be automated.

2. Smart Reconciliation

Integrate AI directly into your accounting system (QuickBooks, Xero, NetSuite, whatever you currently use) to flag discrepancies in real time, not three weeks later when you're frantically trying to close the books for month-end reporting.

Leadership move: Set the acceptable error rate together with your team. Let them define what "flagged for review" actually means in practice. They'll set way higher standards than you would anyway.

3. Weekly AI Financial Summaries

Stop waiting for your accountant to tell you what happened last month when it's already too late to do anything about it. Get insights now while they're still actionable.

> **PROMPT EXAMPLE:** "Summarize this month's financial data and highlight spending trends, anomalies, and category shifts that require leadership attention or decisions."

Leadership move: Share these AI-generated summaries in your regular leadership meetings. Show your executive team you trust AI insights enough to make actual business decisions from them.

4. Human Oversight

AI does the grunt work. Your team does the thinking work. That's the division of labor that actually makes sense in 2025.

Leadership move: Create a crystal-clear approval matrix everyone can see. What can AI do completely autonomously without review? What requires human review before it's final? What requires senior-level sign-off? Write it down. Post it somewhere visible.

Metric Dashboard Visual

Quick Reflection Exercise for Leaders

Which 3 financial processes drain the absolute most time from your team right now?

Which would cause absolutely zero risk if automated first?

Have you actually asked your finance team what they want to automate, or are you just assuming you know?

Set one specific automation goal for the next 30 days, and involve your team in choosing it.

Workflow Two: Operations

Seven Hours of Onboarding from Dusty SOPs

Operations are where most businesses store their hidden gold, buried deep in SOPs collecting digital dust in forgotten folders nobody's opened since 2019. These documents represent literal years of institutional knowledge, best practices, hard-won lessons, and expensive mistakes you never want to repeat. And nobody reads them. Ever.

With AI, those forgotten documents can become your biggest training asset literally overnight.

But here's the brutal leadership reality nobody wants to admit out loud: if your team doesn't use your SOPs right now, AI won't magically make them care. You have to fundamentally change the culture around documentation first, then use AI to make good documentation genuinely great.

How to Lead Operational Systemization:

1. Acknowledge that your current SOPs are probably terrible.

Most SOPs are written by people who know the process so intimately they skip obvious steps everyone else needs. Or they're written in dense corporate language that nobody actually talks like in real life. Or they're 40 pages long when they should be maybe four pages max. Or all three at once.

Start by admitting out loud: "Our SOPs aren't working. I know it. You know it. Let's fix them together."

I had to do exactly this in late 2022. Our SOPs were an absolute joke. Scattered across Google Drive, Dropbox, old shared drives, people's personal email folders from three jobs ago. Half were completely outdated. A third were written in technical jargon that even I couldn't understand without a translator. We basically had to start over from absolute scratch.

2. Make documentation a team sport.

Don't assign one poor soul to "update all the SOPs" as some kind of punishment project or performance improvement plan. That's a death sentence for both the person and the project. They'll hate you forever, the project will drag on for six months, and the SOPs still won't be useful at the end.

Instead, ask each team to document just one critical process per month. That's it. Use AI to help them convert their tribal knowledge into clear, actually usable modules. Then share those modules across the entire organization so everyone benefits immediately.

3. Reward people for using and improving documentation.

If you want people to actually value your SOPs, you have to visibly value their engagement with them. Otherwise it's just another compliance checkbox nobody genuinely cares about.

Celebrate loudly when someone:

- Uses an SOP to solve a problem independently without asking for help
- Improves an existing SOP based on real-world experience
- Creates a brand new SOP for a previously undocumented process nobody wrote down
- Make documentation part of your actual culture and your performance reviews, not just a compliance exercise that HR makes you do once a year because some auditor said so.

Case Study: From 2,000 SOPs to Interactive Onboarding

We had more than two thousand SOPs scattered across literally every digital platform known to humanity. Google Drive. Dropbox. SharePoint. Email attachments from 2019. Someone's personal Evernote. No one had any time to read them all. New hires got completely overwhelmed on day one. Managers got increasingly frustrated trying to train people. Training felt like throwing a phone book at someone and hoping they'd somehow figure out how to do their job through osmosis.

With AI (Claude and ChatGPT, mostly, though we tested others), we converted those massive SOPs into digestible 5–7-minute interactive training modules. Each one had a quiz at the end and a quick reference guide. Interactive. Digestible. Actually useful for once in their existence.

The result? Our new hires completed full onboarding in seven hours instead of seven full days, and we didn't lose any quality whatsoever. Actually, comprehension improved dramatically because people could learn at their own pace and test their understanding immediately instead of sitting in a conference room for eight hours straight pretending to pay attention.

But the way bigger win was cultural. Our team stopped seeing SOPs as bureaucratic busywork that legal makes us do for CYA reasons and started seeing them as genuine tools that actually made their jobs easier and helped them succeed faster. When documentation genuinely helps people do their job better, they use it enthusiastically. Revolutionary concept.

Leadership Lesson: Don't automate bad processes. Fix the process first, make it actually useful to real humans, then use AI to scale it across your entire organization efficiently.

Playbook: Operations Workflow

1. Gather All SOPs

Collect them into a single folder or shared drive so you can actually find them when you need them. If you can't find your own SOPs, neither can anyone else on your team.

Leadership move: Assign a cross-functional team to audit what documentation actually exists vs. what should exist based on how work really gets done daily. You'll be genuinely surprised by the massive gaps between what's documented and what people actually do.

> **IDEA ZONE**
>
> AI can turn every dusty document into living, breathing knowledge that people actually want to use instead of avoiding. Ask yourself:
>
> "What processes are we still teaching manually through shadowing that could be learned from an interactive module?"

2. Chunk Content

Use AI to break those 40-page monstrosities into short, focused modules that take 5-10 minutes maximum to complete. Nobody wants to read a 40-page manual ever. Everyone will watch a 5-minute video or read a one-page guide with screenshots.

Leadership move: Test the modules on brand new hires first before rolling them out company-wide. They'll tell you immediately and brutally honestly if something is confusing, missing critical steps, or just doesn't make any sense.

3. Build Your Learning Hub

Host modules in Notion, Google Sites, or an actual learning management system if you have budget for it. Make it accessible from anywhere. Make it searchable. Make it something people actually want to use instead of actively dread using.

Leadership move: Budget for this properly. A decent learning management system isn't an expense, it's a legitimate investment in scalability and reducing your training costs dramatically long-term.

Visual Artifact: SOP-to-Module Template

Choose one critical SOP this week that new hires consistently struggle with. Ask AI to summarize it into 5 bullet points and create 3 quiz questions.

Test it with a frontline employee who's completely unfamiliar with the process. Actually watch them go through it in real time.

Compare their actual comprehension between the original 15-page SOP and the AI-assisted version.

Notice the massive difference when people can actually use what you've given them instead of feeling completely overwhelmed.

Workflow Three: Marketing

30 Days of Content in 30 Minutes

Know this: Regardless of how far along AI has come and will continue to come, marketing is complicated... You need to be asking the folks responsible for your marketing what they are doing to capitalize on the new search rules, etc. It's also outwardly obvious if you were using AI in the wrong way. Do not create AI ".Slop", you will be discovered.

Once your money and operations are automated and running smoothly, marketing becomes the genuinely fun part. AI gives you both scale and creativity at the exact same time. You're no longer choosing between quality and quantity like it's some kind of Sophie's choice. You get both.

But here's where leadership gets really, really tricky: marketing is where your AI-generated content is most visible to your customers, prospects, and competitors. And if it's bad, generic, or sounds like a robot wrote it? Everyone knows immediately.

How to Lead Marketing Automation:

1. Protect your brand voice like it's oxygen for your business.

AI can generate content at massive scale. Like, ridiculous scale. But it absolutely cannot generate your unique voice without extremely clear, detailed direction. If you don't define your brand voice explicitly with specific examples, AI will default to generic corporate speak that sounds exactly like every other company in your industry.

Before you automate a single piece of marketing, sit down and document:

- Your brand personality (how do you actually sound when you're being authentic?)
- Your brand values (what do you genuinely stand for, not what's on your website footer?)
- Your brand guardrails (what would you absolutely never say under any circumstances?)
- Your brand examples (what does truly great content look like when it comes from you?)

I spent two full weeks just documenting our brand voice before we automated our first marketing workflow. Best two weeks I ever invested. Saved us literal months of fixing generic content later.

2. Treat AI as your junior copywriter, not your creative director.

AI drafts. Humans refine and add soul. That's the workflow. Don't ever flip that around.

Your marketing team should spend way less time staring at blank pages suffering from writer's block and way more time editing, improving, and adding the human touches that actually make content connect emotionally with real people.

3. Build quality gates, not just production schedules.

Speed without quality is just noise that damages your brand reputation. Make absolutely sure every single piece of AI-generated content goes through actual human review before it goes live anywhere public-facing.

Create a simple but mandatory checklist:

- Does this genuinely sound like us?
- Does this serve our audience or just fill space on the calendar?
- Does this meet our actual quality standards?
- Would we be genuinely proud to put our name on this?

If the honest answer to any of those is no, don't publish it. No exceptions ever.

Case Study: Prompt → Calendar → Canva Pipeline

We start with one single well-crafted prompt like this:

"Create 30 days of social content for my company targeting small business owners in Texas preparing for severe weather. Focus on IT disaster recovery, data backup best practices, and cybersecurity tips. Tone should be helpful and trustworthy, not salesy or fear-mongering."

AI generates a complete content calendar with topics, captions, hashtags, and specific visual ideas for each individual post. We plug that directly into Canva, apply our brand kit (colors, fonts, logos, the whole thing), and schedule everything. A full month of genuinely professional content in about 30 minutes of actual work.

Not generic content. Not robotic content that sounds like ChatGPT having a stroke. Content that sounds exactly like us, serves our audience with real value, and drives actual measurable results.

But here's what makes it actually work: we've trained our marketing team to edit all AI output through our brand lens. They know instinctively what to keep, what to change completely, and what to scrap entirely and start over. AI gives us the first draft. Our team gives it personality, soul, and our unique voice.

Leadership Lesson: The quality of your AI output is directly proportional to the quality of your input. Garbage prompts with no context create garbage content. Detailed prompts with clear brand guidelines create content you're actually proud of.

Playbook: Marketing Workflow

1. Define Brand Voice

Use AI to help document your tone, personality, and guardrails in excruciating detail. If you don't know exactly what your brand sounds like, AI will make something up. And you absolutely won't like what it creates.

Leadership move: Run a brand voice workshop with your entire marketing team. Collect 10-15 examples of content that sounds perfectly like you and 10-15 examples that definitely don't. Use those specific examples to train AI on your voice.

IDEA ZONE

You don't need more ideas flooding your brain constantly. You need a systematic process for generating, organizing, and executing ideas consistently. Let AI handle the planning and ideation grunt work, and save your human creativity for storytelling and emotional connection.

2. Create Review Gates

AI drafts. Humans edit thoughtfully. Then schedule. This workflow keeps quality consistently high and speed consistently higher than manual creation could ever be.

Leadership move: Assign crystal-clear ownership. Who reviews what content types? Who has final approval authority? Who's personally accountable for maintaining brand consistency? Write it down where everyone can see it.

3. Build Prompt Library

Maintain a growing library of reusable prompts organized by campaign type. The more specific and detailed your prompts, the better your output will be every single time.

Leadership move: Treat your prompt library like valuable intellectual property. Protect it. Continuously refine it based on actual results. Share it only with trained team members who understand how to use it properly.

4. Measure & Adjust

AI tools can automatically report engagement trends and performance data. Use that data to continuously improve what you're creating, not just to prove you posted something this week to your boss.

Leadership move: Review content performance monthly at minimum. What content is genuinely resonating with your audience? What's completely falling flat? Adjust your prompts and overall strategy based on real performance data, not gut feelings.

PROMPT EXAMPLE: "Generate 10 social post ideas based on our service values of trust, transparency, and speed."

Advanced Leadership Prompt (way, way better):

"As an expert marketing copywriter, Generate 10 social post ideas that position us as the expert choice for homeowners in Colorado dealing with HVAC and plumbing issues. Each post should clearly reflect our brand values of trust, transparency, and speed. Speak directly to the pain point of emergency home repairs. End each post with a specific call-to-action. Tone should be confident, helpful, and reassuring without being salesy or pushy. Include a mix of educational, behind-the-scenes, and customer story formats."

See the massive difference? Specificity and context create genuinely good content instead of generic garbage.

Visual Artifact: 30-Day Content Table

- CTA Library Visual
- Book your free consultation today
- Learn how we save homeowners time with AI
- Follow for weekly home maintenance insights
- Try this tip in your business today
- Download our free winter prep checklist

Content Challenge for Leaders

- Use AI to generate 10 post ideas for your specific business. Pick your top three based on what would resonate most with your actual audience.
- Have three different team members independently edit those same ideas through your brand lens.
- Compare all the variations side by side. Which one sounds most authentically like your brand?
- That person should be the one training everyone else on brand voice and content standards going forward.

Bringing It All Together

When you systematically automate your Finance, Operations, and Marketing workflows, you don't just save some time on your calendar. You fundamentally shift your entire business culture from constantly reactive to genuinely strategic. From buried under daily busywork to actually focused on what matters most for real growth.

But automation without strong leadership is just chaos with fancier tools and bigger problems down the road.

You need real systemization. You need clear, documented processes everyone can follow. You need workflows that make sense. You need quality standards people actually understand. You need feedback loops that work in reality. You need your team to deeply understand not just how to use these AI tools, but why it matters to them personally and where they still add absolutely irreplaceable human value.

These three workflows are your launch pads. They build unstoppable momentum. They prove to your skeptical team that AI isn't something to fear or actively resist. It's something genuinely useful. Something that makes their jobs measurably better, not obsolete or automated away.

Here's the uncomfortable truth: your competitors are already doing this right now. The ones who aren't are falling further behind every single week. The real question isn't whether to systemize your business with AI anymore. The question is whether you're going to actively lead this transformation or passively watch it happen to you while you fall behind.

The 90-Day Leadership Action Plan

Want to know exactly how to roll this out successfully without it being a total disaster? Here's the detailed roadmap I've used with multiple companies ranging from 15 to 200 employees.

Days 1-30: Foundation & Pilot

- Week 1: Audit current processes across Finance, Operations, and Marketing (be brutally honest, not optimistic)
- Week 2: Select your pilot team of 3-5 AI champions (curious, influential, tech-comfortable)
- Week 3: Train pilot team intensively and identify first automation targets together
- Week 4: Launch pilot projects and start documenting early wins and inevitable failures

Days 31-60: Expansion & Systemization

- Week 5: Share pilot results with department leads (celebrate wins loudly, discuss failures openly)
- Week 6: Train Wave 2 consisting of managers and department leads

- Week 7: Build comprehensive prompt libraries and detailed process documentation
- Week 8: Roll out automation to your first complete department

Days 61-90: Scaling & Optimization

- Week 9: Train Wave 3 with frontline teams, tailored specifically by department and role
- Week 10: Implement robust feedback loops and clear quality gates
- Week 11: Measure actual impact and adjust processes based on real data
- Week 12: Celebrate all wins publicly and plan your next-phase automation

Leadership Check-In Questions:

At the end of each 30-day phase, honestly ask yourself:

- Am I personally modeling AI adoption in my own daily work?
- Do my people genuinely trust this process or just comply?
- Are we measuring things that actually matter to the business?
- What's working well enough that we should scale it immediately?
- What's clearly failing that we should fix quickly or abandon entirely?

Exercise: Pick One and Begin

Choose just one workflow today. Right now. Identify a single task that's genuinely repetitive, easily measurable, and extremely low risk if something goes wrong.

Ask AI to create a complete process or detailed prompt for it. Track your specific before-and-after results with real numbers, not estimates.

But here's the critical leadership piece: don't just implement it yourself and call it done. Teach someone else on your team to do it. That's how you actually know if it's truly systemized or if it just lives in your head.

That's how AI goes from interesting theory you read about to real business transformation you experience. That's how you prove it actually works before you scale it company-wide. And that's how you build a thriving culture where AI isn't something that happened to your business from the outside, it's something your team actively chose to embrace because it genuinely made their work better.

The Final Word on Leadership

You can have the most brilliant AI strategy in the entire world, backed by unlimited budget and the absolute best tools money can buy. But if you can't get your actual people to adopt it and use it consistently, it's completely worthless. Total waste of money and time.

Leadership in the AI era isn't about having all the right answers on day one. Nobody does. It's about asking the right questions. Creating the right environment where experimentation is genuinely safe. And showing your team through your own actions that change is possible and absolutely worth the temporary discomfort.

When you automate where it hurts your business first, you don't just reclaim hours on everyone's calendar. You reclaim confidence. You reclaim clarity. You reclaim focus on what actually matters instead of busywork. You prove to your entire team that AI isn't some existential threat to their careers, it's a powerful tool that makes them way better at their jobs.

That's the kind of competitive advantage that compounds over time. That's the kind of transformational leadership that builds organizations people actually want to work for long-term.

And that's exactly where your AI journey truly begins.

CHAPTER 4

Design : Safety, Policy, and People

When Power Meets Risk

"It would be foolish to think that with all of this AI:where hackers have unlimited power for $20 a month:there wouldn't be some drawbacks to implementing AI in your organization."

Safety and security are not theoretical concerns; they're very real, and they can take down even the most innovative teams if ignored.

So how do you protect yourself?

One of the things I tell audiences all the time is that you must be super diligent about connectors. How are your employees allowed to connect to your internal databases, shared drives, CRMs, or project folders? These connectors make life incredibly easy when you're pulling data for analytics, reports, or planning:all the great reasons we love AI.

But here's the problem: every new connection is a new doorway. And if those doors aren't locked, your data could be walking right out into the world. That's a nightmare scenario if you handle sensitive client information.

You need to get with your IT team right away and make sure they understand what tools people are using, what's being connected, and what policies are in place.

And let's talk about "free."

The free versions of most AI tools offer zero protection. They are training on your data, they are open to the world, and they create a massive vulnerability in your organization: for what? To save twenty dollars a month? That trade-off doesn't make sense anymore.

Safety as Strategy

Safety isn't a brake pedal:it's the steering wheel.

When you build your AI systems with safety and policy in mind from the beginning, you don't slow innovation; you guide it.

In my company, Denver Plumber & HVAC, we implemented AI to streamline customer service and dispatching. But before we let a single bot talk to a customer, we had a "safety check" conversation.

What data would it see? Where would it store responses? Who could access the logs?

That one conversation saved us from exposing customer home addresses and service notes through an AI workflow.

And the best part? Once everyone knew the boundaries, the team moved faster because they weren't afraid to break something.

> "Design like someone's job depends on it:because it probably does."
>
> When people feel safe using AI, they experiment more confidently.
>
> That's the paradox of protection: the stronger your guardrails, the more freely your team can innovate.

Data Protection : What's Yours, What's Not

You can't protect what you don't understand.

Start by mapping your organization's data.

Three Data Categories to Define:

- Sensitive: client information, financials, personal data
- Internal: training docs, playbooks, policies
- Public: marketing materials, blogs, press releases

Once you see the landscape, you can decide what's safe to use in AI tools:and what's absolutely off-limits.

Rule of Thumb: If you wouldn't email it to a stranger, don't feed it to a public AI model.

Data Audit Sprint

Gather your leadership team.

In 30 minutes, list every system that holds key data (Dropbox, CRM, spreadsheets, Slack, etc.).

Classify each as Sensitive / Internal / Public.

Discuss where AI fits:and where it shouldn't.

You'll be surprised what you uncover.

Tool Selection : Enterprise vs. Personal Use

Here's where a lot of teams get tripped up: they want to use AI but don't want to pay for enterprise accounts.

That's like building a mansion and locking it with a luggage padlock.

Enterprise-level AI tools:like ChatGPT Team, Google Gemini, Copilot for Microsoft 365, or Anthropic Claude for Business:come with encryption, data isolation, and admin control.

Personal accounts: especially the free ones: do not.

"Saving $20 could cost you $20 million."

If your company handles any customer data:names, addresses, financial info: you owe it to your clients and yourself to use secure AI platforms.

Review Gates : Safety Stops That Build Speed (The Human in the Loop)

A review gate is a planned checkpoint before or after an AI process goes live. It's how you move fast without losing control.

You can implement these in any organization:

Pre-launch Gate:
Before publishing AI-generated content or reports, someone reviews for accuracy, compliance, and tone.

Post-launch Gate:
Quarterly or biannual reviews to check for drift, data exposure, or new vulnerabilities.

Changing the Talk-Track : From Fear to Empowerment

When people hear "AI policy," they often tense up. They picture red tape, micromanagement:or job loss. That fear is a killer of progress.

If you want adoption, change the talk-track. Replace fear with empowerment.

During one of my sessions with educators, I watched the room shift the moment I reframed AI from "dangerous" to "designed for good."

Once people realized they could shape how AI worked in their world, fear turned into curiosity:and curiosity turned into progress.

"Flip the Script"

Once your people feel agency, they start using AI with purpose: and that's when real transformation begins.

AI Acceptable Use + Vendor Checklist

Your action plan for this chapter.

AI Acceptable Use Policy Template

- Define what's allowed: approved tools, workflows, and data types
- Define what's restricted: sensitive data, external connectors
- Establish accountability: who reviews usage, and how issues are reported
- Schedule reviews: quarterly policy updates and staff acknowledgment

Vendor Evaluation Checklist

Before adopting a new AI product, ask:

- How does this tool store and protect data?
- Is my data used for model training?
- Does it comply with SOC 2, GDPR, or HIPAA?
- Can I delete or export my data anytime?
- Does the vendor publish transparency reports or bias audits?

Treat this checklist like your AI seatbelt.
You might never need it: but you'll be grateful for it when something goes wrong.

Closing: Safety is the New Speed

When you design with safety, you don't slow down innovation:
you make it sustainable.

Safety builds confidence.
Confidence drives adoption.
Adoption creates transformation.

That's the real secret to leading in the age of AI: you're not just protecting your data: you're protecting your people, your reputation, and your ability to grow with confidence.

CHAPTER 5

Deploy:Pilot, Measure, Scale

"It's not real until you deploy."

Every great AI idea shines in the brainstorming stage. The prompts flow, the tools impress, and everyone's talking about transformation. But until you deploy, until someone actually uses the workflow and measures the results, it's still just theory. And theory doesn't show up on your P&L.

Here's the truth: AI isn't a switch you flip. It's a dimmer; you turn up one successful pilot at a time.

This chapter gives you the framework to stop talking about AI and start proving it works. You'll learn how to pilot like a startup, measure what actually matters, and scale what earns the right to survive. No fluff. No endless committees. Just disciplined iteration that turns experiments into competitive advantage.

The Deployment Framework

Deploy = Pilot → Measure → Scale

Think of this as your three-gear engine. Each gear turns the next. Skip one, and the whole system stalls.

1. The Pilot: One Owner, One Metric, 30 Days

You can't "sort of" pilot AI. You either run a structured test with clear accountability, or you drift in Pilot Purgatory, where enthusiasm replaces evidence and nothing ever ships.

I've watched companies spend six months "exploring possibilities" without ever deploying a single workflow. They hold meetings about meetings, create committees to form other committees, and wonder why their competitors are eating their lunch.

Stop doing that.

The Three Golden Rules of Piloting

> *"Speed matters. Thirty days gives you clarity before complexity takes over."*
>
> - SUSAN FREW

These rules aren't suggestions. They're guardrails that keep you honest. The moment you start adding co-owners or tracking seven metrics, you've already lost focus. And without focus, you can't learn what actually works.

PILOT LIKE A STARTUP

Treat each pilot like a mini-startup:

You have a founder. (the pilot owner).
You have an investor (the sponsor or exec approving time and resources).
You have a hypothesis. (the metric you're testing).

At the end of 30 days, ask yourself:

"If this pilot were its own tiny company, would I invest again after 30 days?"

If the answer is no, you kill it. If it's yes, you scale it. If it's maybe, you iterate once and test again. That's it. No endless debates. No "let's give it another quarter." You decide and move.

This mindset shift alone will save you from months of wasted effort.

2. Measure What Matters

Data doesn't have to be fancy. It just has to be focused.

You're not building a perfect analytics platform or impressing investors with real-time dashboards. You're testing if the AI is delivering value. That's it. Can you prove this workflow saves time, reduces errors, or generates revenue? Yes or no?

Weekly Measurement Checklist

Notice what's not here: complex formulas, multiple data sources, or real-time integrations. You're tracking one thing, once a week, with a color code. Green means keep going. Yellow means iterate. Red means kill or pivot.

That's all you need to know if something's working.

DASHBOARDING BASIC Lightbilb

Keep your first AI dashboard brutally simple:

3 to 5 KPIs max

Track what moves the needle. If you can't explain why a metric matters in one sentence, cut it.

Weekly updates (not real-time yet)

Real-time dashboards are seductive, but they distract you from the learning. Weekly is enough to spot trends without drowning in noise.

Color codes:

● Keep (exceeding expectations)

● Iterate (partial success, needs tweaking)

● Kill (not working, cut losses)

Remember: The dashboard isn't for presentation. It's for decision-making. Build it for clarity, not applause.

3. The Kill / Keep / Iterate Framework

At the end of your 30-day pilot, hold a truth session. This is where leadership, data, and intuition meet. No spin. No politics. Just an honest assessment of what happened.

The only bad outcome is indecision.
"Maybe" is not a strategy.

Here's what most companies get wrong: they can't kill a bad pilot. They keep pouring resources into something that clearly isn't working because someone's ego is attached or because "we've already invested so much time."

Stop doing that. Killing a pilot isn't failure. It's learning. Every dead pilot teaches you something valuable about your business DNA: what your team will actually use, what they'll resist, and where the real friction lives.

The goal isn't to have every pilot succeed. The goal is to learn fast and invest your resources where AI actually delivers.

4. Scale: From Pilot to Process

When a pilot earns a Keep, it graduates to a Standard Operating Procedure (SOP). That's when AI shifts from "cool experiment" to competitive advantage.

Scaling isn't about rolling out to everyone at once. It's about turning your successful workflow into a repeatable process that anyone on your team can execute without starting from scratch.

Scaling Checklist

- Document workflow clearly. (step-by-step, no assumptions)
- Create a short Loom or internal video demo. (Show, don't just tell.)
- Add to company playbook or Notion SOP. (Make it findable and searchable.)
- Train the next 3 to 5 users. (expand cautiously, not recklessly)
- Automate repetitive prompts or create an AI agent. (reduce friction, increase consistency)

Each successful pilot becomes a building block. After a few months, you're not just running isolated experiments anymore. You're building an internal AI playbook that compounds value with every deployment.

Transformation Callout

"When three pilots turn into company-wide practices, AI stops being a side project and becomes your operating system."

That's the shift you're aiming for. AI isn't a department. It's not a special initiative. It's woven into how your business operates every single day.

Artifact: 30-Day Pilot Charter

The Charter keeps your pilots disciplined and visible. Use this every time you test a new AI workflow. It forces clarity before you start and accountability when you finish.

30-Day AI Pilot Charter Template

Fill this out before you launch. Review it every week. Make your final call at the end of 30 days. No exceptions.

This simple template will save you from scope creep, vague goals, and the deadly "we'll figure it out as we go" approach that kills most AI initiatives before they ever deliver results.

The Pilot Is The Learning Loop

Each pilot creates a small feedback loop that accelerates learning. The more loops you run, the faster your organization evolves.

Think about it this way:

Run 1 pilot per quarter, and by year-end, you'll have 4 proven use cases and a new internal AI playbook.

That's not just progress. That's transformation at a pace your competitors can't match because they're still stuck in planning mode.

Closing Thought

You don't build AI mastery by reading about it. You build it by piloting, measuring, and scaling with intention and accountability.

One owner.
One metric.
Thirty days.

That's how you transform ideas into impact. That's how you stop talking about AI and start proving it works. And that's how you build a competitive advantage that compounds quarter after quarter.

Deploy. Measure. Scale. Repeat.

PART III

Foundation Workflows

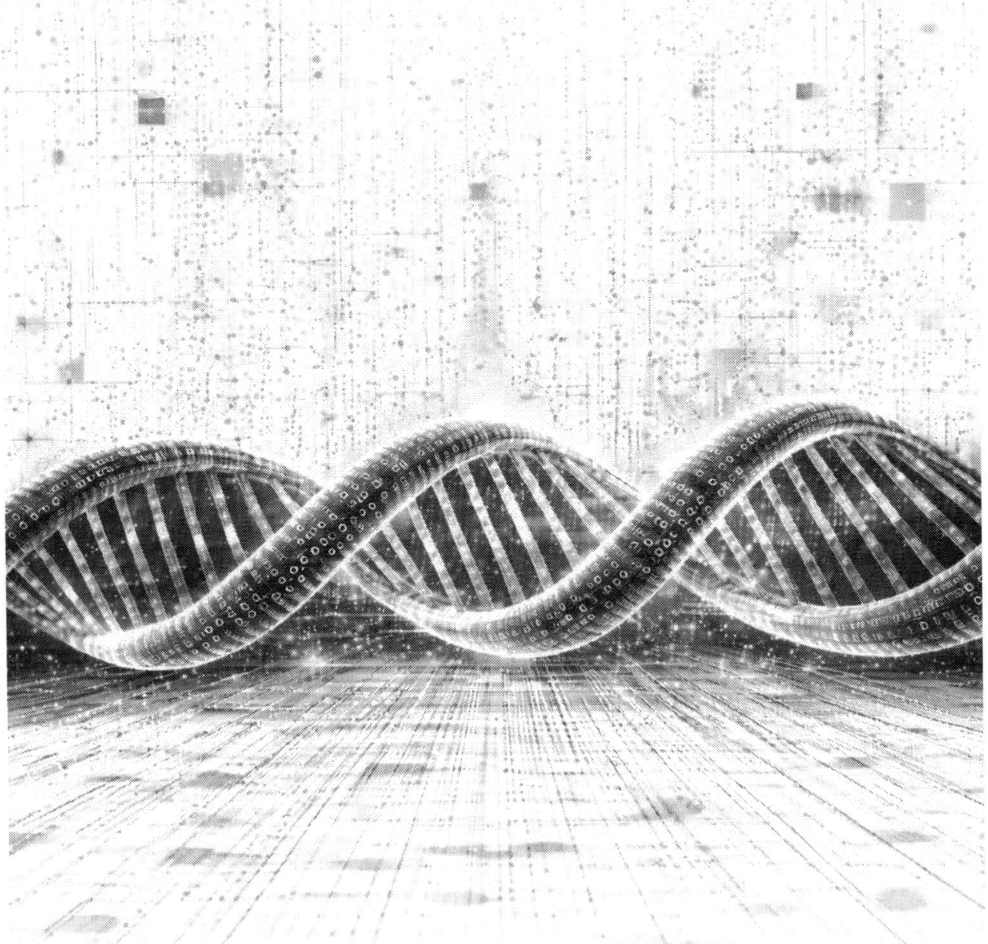

CHAPTER 6

Finance: Cut Waste Without Cutting People

Gear made of coins turning - symbolizing efficient cash flow

AI isn't here to replace your finance team. It's here to give them their time back.

Here's what most business leaders get wrong about AI in finance. They think automation means elimination. Fire the bookkeeper, cut the AP clerk, slash overhead. But that's not transformation. That's just fear with a calculator.

Real transformation? That happens when you stop wasting human intelligence on work a machine can do better. When your finance team spends less time chasing down invoices and more time analyzing what those invoices actually mean for your business. When your CFO can forecast instead of firefight.

Look, the goal isn't to cut headcount. It's to streamline repetitive tasks so people can focus on decision-making, forecasting, and customer value. Because let's be honest here. Your finance pro didn't go to school for years just to manually match line items in a spreadsheet. They're overqualified for the grunt work, and you're overpaying for it.

This chapter shows you exactly where AI belongs in your finance operations, how to implement it without chaos, and how to measure success in ways that matter. Not theory. Real numbers. Real processes. Real results.

CASE STUDY: Automation That Pays for Itself

Let me show you what this looks like in the real world.

Small manufacturing business. Twenty-three employees. Drowning in financial admin.

Their part-time bookkeeper spent over 12 hours every week on invoice matching, vendor communication, and reconciliation. Their office manager? Another six hours fielding calls about billing questions, payment status, and account inquiries.

That's 18 hours weekly. Nearly half a full-time position just managing the paper trail.

The owner knew something had to change. But here's the thing. She refused to lay anyone off. Her team was loyal, skilled, and essential. She just needed them doing different work.

Before AI implementation:

- Twelve hours weekly on manual bookkeeping tasks
- Six hours weekly fielding finance-related calls
- Three to five day invoice processing cycle
- Frequent errors requiring rework
- Team frustration at repetitive work
- Zero bandwidth for financial analysis or planning

After AI implementation:

They started small. Real small. An AI bookkeeping assistant to handle invoice intake and classification. An AI-powered call routing system to triage vendor and customer inquiries. Simple tools. Clear parameters. Human oversight built in from day one.

Sixty days later, here's what happened:

- Bookkeeping time dropped 68% (down to 4 hours weekly)
- Phone hold times cut by 42%
- Invoice cycle time fell to 1.5 days
- Both team members moved to growth projects without a single layoff
- Bookkeeper now focuses on cash flow analysis and vendor negotiations
- Office manager handles customer success instead of billing questions

The financial impact? They saved roughly $28,000 annually in labor costs while increasing team satisfaction and productivity. The bookkeeper actually told me she finally felt like a financial professional instead of a data entry clerk.

"We didn't replace people.
We replaced repetitive."

- SUSAN FREW

That's the difference between cost-cutting and smart operations. One shrinks your capacity. The other expands it.

"But What If..."

"What if AI makes a costly mistake?"

Valid concern. Here's the truth: AI will make mistakes. So do humans. The difference? AI makes consistent, predictable mistakes you can catch with oversight rules. Humans make random, unpredictable ones you might not catch for months.

Solution: Build approval thresholds. Any payment over $500? Human reviews it. Any vendor discrepancy over 10%? Flags for attention. You're not removing oversight. You're making it more efficient.

"Our accounting system is too old for this."

Probably not as old as you think. Most modern AI tools don't require replacing your entire tech stack. They sit on top of what you already have. QuickBooks from 2015? Still works. Excel spreadsheets? AI loves those.

You don't need a new ERP system. You need smart connectors between what you have and what AI can do. Tools like Zapier, Make, or even ChatGPT with API access can bridge that gap for a few hundred bucks a month.

"We tried automation before and it was a disaster."

Yeah, because five years ago "automation" meant rigid, rule-based systems that broke every time something changed. Modern AI is different. It learns. It adapts. It understands context.

The automation that failed you in 2019 isn't the AI available in 2025. Don't let old scars stop you from new opportunities.

"My team will never go for this."

They will if you position it right. Don't tell them you're automating their jobs. Tell them you're automating the parts of their jobs they hate. Nobody loves data entry. Nobody gets excited about matching line items for six hours.

Frame it as giving them their skills back. Most finance people will jump at the chance to do actual financial work instead of administrative drudgery.

"What about compliance and audit trails?"

WHERE AI BELONGS IN FINANCE

Not every finance task needs AI. Some require human judgment, relationship management, and strategic thinking. But plenty of tasks don't.

Start small. Focus on places where humans are overqualified for the work. Where the rules are clear, the inputs are structured, and the output is predictable. That's AI's sweet spot.

High-impact starting points:

Vendor invoice matching - AI reads invoices, extracts data, matches to purchase orders, and flags discrepancies faster than any human. Your team reviews exceptions, not every single line item.

Receipt classification - Stop manually sorting expense receipts into categories. AI reads, categorizes, and organizes based on rules you set once.

Bank reconciliation - Daily transaction matching doesn't require human intelligence. It requires accuracy and consistency, which AI delivers without coffee breaks.

Expense reporting - Employees snap photos. AI extracts data, categorizes, checks policy compliance, and routes for approval. Done.

Budget variance explanations - AI analyzes spending patterns and generates plain-English summaries of why actual spend differs from budget. Your CFO reviews insights, not spreadsheets.

Here's the pattern: AI thrives on rules and repetition. Exactly where your people shouldn't have to live.

Think about it this way. Your finance team has expertise in cash flow strategy, vendor relationships, forecasting, compliance, and risk management. Every hour they spend on data entry? That's an hour stolen from what they're actually good at.

AI doesn't replace judgment. It clears the path so judgment can happen faster and more often.

Most AI finance tools create better audit trails than humans do. Every action is logged. Every decision is documented. Every exception is tracked. Your auditors will actually love this.

Plus, you can configure AI to follow SOX requirements, GAAP standards, or industry-specific regulations. The compliance argument actually favors AI, not against it.

PLAYBOOK: Automate the Financial Flow

Here's your step-by-step implementation path for AI-enhanced finance operations. This isn't a rip-and-replace strategy. It's gradual automation that builds confidence while delivering results.

Step 1: Invoices - Intake & OCR

Use AI to scan, read, and categorize invoices automatically. No more manual data entry from PDFs or paper documents.

How it works: Vendors email invoices to a dedicated address. AI extracts vendor name, invoice number, date, line items, totals, and payment terms. Data flows directly into your accounting system.

Tools to consider: Many modern accounting platforms have this built in. Otherwise, tools like Dext, Hubdoc, or AI-powered integrations with QuickBooks or Xero handle this beautifully.

What stays human: Setting up vendor rules, reviewing flagged exceptions, approving payments over certain thresholds.

Step 2: Matching - Rules Engine + LLM Cross-Check

Auto-match invoice line items to purchase orders and receipts. Flag mismatches for human review.

How it works: AI compares invoice details against your PO database and received goods records. When everything aligns, it moves forward. When something's off (wrong price, quantity mismatch, unexpected charge), it flags for review.

The intelligence layer: Modern AI understands context. "Rush shipping" on an invoice can be matched to "expedited delivery" on a PO. It learns your vendor language patterns.

What stays human: Reviewing flagged items, approving exceptions, updating vendor terms.

Step 3: Reconciliation - Daily Digest

Use AI to summarize exceptions, mismatches, and balances in human language. No more hunting through transaction lists.

How it works: Each morning, your finance team gets a plain-English summary. "Twenty-three transactions matched automatically. Three require review: two potential duplicates, one pricing variance with XYZ Supplier."

The time savings: Instead of reviewing 100% of transactions, your team focuses on the 5-10% that need attention.

What stays human: Final reconciliation approval, investigating patterns of errors, updating automation rules.

Step 4: Call Routing - Smart Front Door

Deploy AI at your phone or chat front-end to triage billing and vendor calls. No hold music needed.

How it works: AI answers common questions instantly. "When will my payment be processed?" "What's my account balance?" "How do I update my billing address?" Complex issues route to the right human immediately.

Real impact: One client cut average call handle time from eight minutes to three minutes. Not because people talked faster. Because simple questions never reached a human.

What stays human: Disputes, relationship management, negotiation, complex problem-solving.

Step 5: Handoff - Human Oversight

Final checks by your finance lead ensure compliance and accuracy. AI processes, humans approve.

How it works: Build approval workflows based on dollar amounts, vendor type, or transaction category. AI handles the heavy lifting. Humans make the calls.

The safety net: You're not surrendering control. You're automating the routine so you can focus on the strategic.

What stays human: Everything that matters. Vendor relationships. Payment prioritization. Cash flow decisions. Compliance verification.

The key principle: Each step reduces manual work while increasing accuracy. Your team doesn't work less. They work smarter on things that actually move the business forward.

AI Finance Tools Starter Kit

You don't need a dozen tools. You need the right three to five that actually solve your specific problems. Here's what works for most businesses.

For Invoice Processing & OCR:

Dext (formerly Receipt Bank) - Scans invoices and receipts, extracts data, pushes to your accounting system. Works with QuickBooks, Xero, Sage. Around $40/ month for small businesses.

Hubdoc - Similar to Dext but owned by Xero. Great if you're already in that ecosystem. Automated fetch from vendors' websites.

For Bookkeeping Assistance:

ChatGPT Team or Claude Projects - Seriously. Set up a project for your finance workflows with your chart of accounts, categorization rules, and vendor preferences. Use it to draft reconciliation notes, explain variances, and answer routine questions. $30/month.

Botkeeper - Full AI bookkeeping service if you want more hands-off support. They combine AI with human oversight. More expensive but comprehensive.

For Workflow Automation:

Zapier - Connects your accounting software to email, Slack, Google Sheets, and hundreds of other apps. Automate invoice routing, payment notifications, exception alerts. Starts free, scales with usage.

Make (formerly Integromat) - Similar to Zapier but more powerful for complex workflows. Slightly steeper learning curve but worth it for multi-step processes.

For Call & Communication Handling:

ChatGPT with Voice - Can handle inbound vendor questions via phone or chat. Integrate with your phone system or website chat.

Intercom or Drift - Business chat platforms with AI features built in. Good for vendor portals and customer billing inquiries.

For Expense Management:

Expensify - AI reads receipts, categorizes expenses, enforces policy, routes for approval. Integrates with most accounting platforms. Great for teams with travel or field expenses.

> **PRO TIP:** Don't buy them all at once. Start with invoice processing (biggest time suck), then add workflow automation, then layer in communication tools. Each tool should solve a specific pain point you've already identified.

Budget reality check: You can get started with meaningful AI finance automation for under $200/month. Compare that to the cost of even 10 hours of bookkeeping time. ROI happens fast.

METRIC PACK: Finance Performance Tracker

Cost per Doc | Cycle Time | Exception Rate

You can't manage what you don't measure. These three metrics tell you if your AI implementation is actually working or just adding complexity.

> Tie these metrics to quarterly OKRs. Makes AI progress measurable and rewardable. When your finance team sees their efficiency gains in actual numbers, they become AI champions instead of skeptics.

How to track these:

Start with a baseline week before any AI implementation. Track manually if you need to. You need real numbers, not estimates.

Then measure weekly for the first 30 days. Bi-weekly for the next 60 days. Monthly after that.

What good looks like:

- Cost per document drops steadily as automation handles more volume
- Cycle time decreases without sacrificing accuracy
- Exception rate starts high (maybe 10-15%) as AI learns, then drops below 5% as rules refine

What failure looks like:

- Cost per document stays flat or rises (you're adding technology without removing work)

- Cycle time increases (automation is creating bottlenecks, not removing them)
- Exception rate stays above 10% (AI isn't learning your patterns, or your processes are too inconsistent)

BONUS MOVE: Share wins publicly. "Team cut invoice processing time by 45% this quarter" sounds a lot better than "we implemented new software."

RED FLAGS: When AI Finance Goes Wrong

Not all AI implementations succeed. Here's how to spot trouble early and course-correct before small problems become expensive disasters.

Red Flag #1: Exception Rate Stays Above 15%

If three weeks in, you're still manually intervening on more than 15% of transactions, something's wrong. Either your AI rules are too loose, your vendor data is too inconsistent, or your processes need cleaning up before automation.

What to do: Pause expansion. Analyze the exceptions. Are they random or patterned? If patterned, fix the root cause. If random, tighten your automation parameters.

Red Flag #2: Team Resistance Is Growing, Not Shrinking

Week one resistance is normal. Week four resistance is a problem. If your finance team is finding workarounds to avoid using the AI tools, you've got a people problem disguised as a technology problem.

What to do: Talk to them. Really listen. Often the issue isn't the AI itself but how it was introduced, lack of training, or fear about job security. Address the human concerns before tweaking the tech.

Red Flag #3: Accuracy Is Dropping

AI should maintain or improve accuracy, not reduce it. If you're seeing more payment errors, more reconciliation mismatches, or more vendor complaints, stop immediately.

What to do: Audit the last 50 transactions manually. Find where accuracy broke down. Was it data quality issues? Poor rule configuration? Insufficient human oversight? Fix the cause, not the symptom.

Red Flag #4: You Can't Explain How It Works

If the person running your AI finance system can't explain to an auditor or CFO exactly how invoices are processed, matched, and approved, you've got a compliance time bomb.

What to do: Document everything. Create process maps. Write down approval rules. AI should make your processes more transparent, not more mysterious.

Red Flag #5: Cost Savings Aren't Materializing

If you're three months in and haven't seen measurable reduction in labor hours or processing costs, the automation isn't working. You're just adding technology on top of existing inefficiency.

What to do: Go back to your baseline metrics. Where was the promised time savings supposed to come from? If it's not happening, either the automation isn't being used consistently or you automated the wrong things.

The bottom line: AI finance automation should feel like relief, not stress. If your team is more frustrated, your costs are higher, or your accuracy is worse, stop and reassess. Done right, the wins show up fast and compound over time.

COMPLIANCE NOTE: AI & Financial Regulations

Let's address the elephant in the room. Can AI tools meet SOX requirements, GAAP standards, and industry-specific financial regulations?

Short answer: Yes. When configured properly.

Here's what regulators and auditors actually care about: consistent processes, complete audit trails, segregation of duties, and documented controls. AI can deliver all of that, often better than manual processes.

The Human Dividend

Every hour AI gives back to your finance team is an hour they can spend forecasting, analyzing, and innovating.

Think about what your finance people could do with an extra ten hours per week. Not hypothetically. Actually think about it.

They could build better cash flow models. Negotiate better vendor terms. Identify cost-saving opportunities before they become problems. Analyze spending patterns to inform strategic decisions. Train other departments on financial literacy. Build relationships with key vendors.

All the things they went to school for. All the things you hired them to do.

You're not cutting people. You're compounding potential.

That's the human dividend of AI. When you stop treating talented people like data processors, they start acting like strategic partners. Clear the repetitive work, and space opens for creative problem-solving.

And here's what nobody tells you. People want their jobs to matter. They want to use their brains for more than matching numbers. Give them that opportunity and watch what happens.

I've seen bookkeepers become financial strategists. AP clerks become vendor relationship managers. Finance admins become process optimization experts.

Same people. Different work. Bigger impact.

That's not replacement. That's transformation.

Visual Summary: Finance Workflow Map

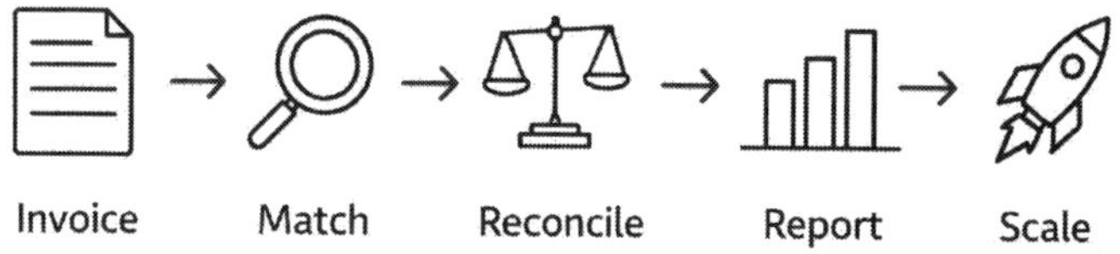

Each step has a small icon representing the automation layer with a human oversight checkpoint.

Caption: "When numbers flow, so does focus."

AI in finance isn't about cutting corners. It's about cutting waste so your team can focus on what actually creates value.

You don't need a massive budget or a technology overhaul. You need clarity on where human intelligence is being wasted. And the willingness to let AI handle the repetitive while your people handle the strategic.

Start with one workflow. Measure the impact. Scale what works.

Your finance team will thank you. Your cash flow will reflect it. And your business will run smoother than it ever has.

Next up: How to apply this same thinking to operations without turning your team into robots.

CHAPTER 7

Operations : 7 Hours of Onboarding from Dusty SOPs

Let me tell you about the most useless thing in your company right now.

It's not the broken coffee maker in the break room or that one employee who still uses Internet Explorer. It's your Standard Operating Procedures. You know, that massive collection of documents you spent months creating, carefully organizing into folders, and then promptly forgot existed.

I know this because I've seen it in every company I've worked with. During my years running companies, we had SOPs for everything. We documented it all like good little business owners. And you know what happened? Nobody read them. Not one person. New employees would get handed a binder on their first day, nod politely, and then spend the next three weeks asking their coworkers how things actually worked around here.

Sound familiar?

Here's what nobody tells you about SOPs. The problem isn't that you created them wrong. The problem is that we're still treating knowledge transfer like it's 1997. We dump information into Word documents, file them in a shared drive, and expect human beings to magically absorb dense procedural text on their own time.

Spoiler alert: they don't.

But here's what got me excited about AI in the first place. What if those dusty SOPs could actually talk back? What if they could quiz someone to make sure they got it? What if institutional knowledge didn't live in some forgotten folder but showed up like a patient teacher who answered questions at 2 AM when no manager was awake?

That's not science fiction. Companies across industries are doing exactly this right now. And the results aren't small tweaks. We're talking about cutting onboarding time in half while actually improving how well people retain information.

Let me show you how real companies are making this work, then I'll walk you through the exact process so you can do it too.

CASE STUDY: From 2,000 SOPs to an AI Onboarding System

A mid-sized healthcare services company had a problem that's probably familiar to you. Over fifteen years, they'd accumulated more than 2,000 pages of standard operating procedures. Not an exaggeration. Two thousand pages.

Patient intake protocols. Insurance verification scripts. Equipment sterilization guides. HIPAA compliance procedures. Emergency response documentation. All the stuff that makes a healthcare business run properly but that nobody has time to actually read.

Their onboarding process was crushing them. New medical assistants and administrative staff had to sit through fourteen hours of reading before they could start working. Fourteen hours in a windowless conference room with highlighters and binders, trying to memorize procedures they'd forget by next Tuesday.

The numbers were brutal. They were losing 40% of new hires within the first 90 days because people felt overwhelmed and unprepared. Managers knew the training didn't work, but rebuilding it from scratch? With their workload, that wasn't happening.

So we tried something different. We took those 2,000 pages and fed them into an AI system. Then we let the AI do what no human trainer has time to do: break everything into bite-sized modules, organize them by role and skill level, create comprehension checks, and build a searchable knowledge base that people could actually navigate.

The AI didn't just summarize the content. Think bigger than that. It restructured the entire learning experience. Instead of "Read Section 4.2.7 on Patient Intake Procedures," new hires got a five-minute module that walked them through each step, explained why it mattered for patient safety, and tested their understanding before they moved forward.

They built the whole thing using tools they already had. No expensive LMS platform. No developers. Just AI doing the heavy lifting of content transformation, Notion as a delivery system, and a simple dashboard to track who was actually learning what.

The results were remarkable:

Onboarding time dropped from 14 hours to 7 hours. New staff were cleared to work with patients in half the time without compromising on comprehension or compliance.

90-day retention improved by 38%. When people understood their role from day one and had on-demand access to procedures, they were far less likely to quit feeling overwhelmed.

The knowledge base logged over 400 searches per week. Staff were actually using the system to find answers instead of interrupting their supervisors constantly.

Their training director told me something I'll never forget: "For the first time in my career, I'm not re-explaining the same HIPAA procedures every single week. The system does it for me consistently. And when compliance requirements change, I update one module and everyone gets retrained automatically."

I've seen similar results across industries. A property management company cut their maintenance tech onboarding from 16 hours to 6 hours and saw quiz comprehension scores jump to 87%. A financial services firm turned 800 pages of compliance documentation into 45 interactive modules and reduced regulatory violations by new employees by 62% in the first quarter.

The pattern is consistent: when you transform static documents into interactive learning, people actually learn. And when people learn properly from day one, everything else gets easier.

PLAYBOOK: Turn SOPs into Smart Training

Alright, enough theory. Here's exactly how to take your existing procedures and transform them into an onboarding system that actually works. You don't need a development team. You don't need expensive software. You just need a structured approach and the willingness to let AI do the heavy lifting.

Step 1: Gather & Clean

Start by collecting everything. And I mean everything. Pull together all your SOPs, training manuals, policy documents, FAQs, process guides, and any other instructional materials scattered across your organization.

Don't worry if they're messy. Most companies have SOPs written by different people over different years in completely different formats. Some are two-page checklists. Others are fifty-page technical manuals with screenshots from Windows XP. Some are beautifully organized. Others barely make sense.

Gather it all anyway. Then do a quick cleanup pass. Remove obviously outdated content. Flag anything that contradicts current policy. You're not rewriting here. You're just making sure you're not training people on procedures you stopped using three years ago.

Time investment: 2-4 hours depending on how organized your files are.

SOPs AS CONVERSATIONS

Think about the best trainer you ever had. What made them great?

They didn't just hand you a manual and disappear. They explained things clearly. They checked to make sure you were following along. When you looked confused, they answered your questions. They adjusted their pace based on whether you were picking things up quickly or needed more time.

AI can do all of that now. Seriously.

When you convert your SOPs into an AI-powered training system, you're not digitizing text. You're creating a virtual coach that can explain policies conversationally instead of drowning people in corporate jargon. It can test comprehension with actual quizzes that measure understanding instead of just box-checking. And it can adapt to each person's learning pace.

Your dusty binder just became a 24/7 trainer who never gets tired, never gets impatient, and never forgets to cover the important stuff.

This isn't about replacing managers. It's about multiplying their effectiveness. Instead of spending hours walking new hires through the same basic material, managers can focus on coaching, mentoring, and handling complex situations that actually require human judgment.

Plus, once you build it, it keeps working. Update your SOPs? The AI updates the training. Add a new policy? It gets incorporated automatically. Your knowledge base becomes a living resource instead of a static document that goes out of date the second you hit save.

Step 2: Chunk & Label

This is where AI earns its keep. Upload your cleaned SOPs and have the AI break them into learning modules. Each module should be five to ten minutes of focused instruction on a single topic or process.

The AI will analyze your content, identify natural breakpoints, and organize everything into a logical sequence. A 200-page employee handbook becomes twenty digestible modules that new hires can work through at their own pace.

Label each module clearly: title, objective, duration, source document, and target audience. A field technician doesn't need the same training as a customer service rep. Tag your modules so people only see what's relevant to their role.

Time investment: 1-2 hours of AI processing plus 2-3 hours of review and labeling.

Step 3: Generate Quizzes

Knowledge without verification is just wishful thinking. You need to know whether people actually understand what they've learned or if they're just clicking "next" to get through faster.

Use AI to create three to five comprehension questions per module. Mix it up: one question that tests basic recall, one that tests process understanding, and one scenario-based question that requires applying the knowledge to a realistic situation.

The AI generates these in seconds. You just review them to make sure they're testing the right concepts and adjust any questions that feel too easy or unnecessarily tricky.

Time investment: 30 minutes per 10 modules.

Step 4: Deliver in an LMS-Lite Tool

You don't need a fancy Learning Management System. Most companies can host training modules in tools they already use.

Notion works great for this. So does ClickUp. Even a well-organized shared drive with embedded AI chat access can function as a training platform.

The key is making the content easy to navigate and giving employees a way to ask questions when they get stuck. Embed an AI chatbot trained on your SOPs, and suddenly your static documents become an interactive resource that can troubleshoot problems in real time.

Time investment: 4-6 hours to set up the initial structure and integrate AI chat.

Step 5: Track Progress

If you're not measuring it, you're not managing it. Set up a simple dashboard to track module completion, quiz scores, and time to competency.

Most tools can export this data automatically. You just decide what metrics matter. Are people finishing modules but failing the quizzes? That's a content problem. Are they acing the tests but taking three times longer than expected? That's a pacing issue.

Use the data to refine your training over time. The beauty of an AI-powered system is that it gets better as you use it.

Time investment: 1 hour to set up tracking, then 30 minutes monthly to review data.

EXERCISE 1: The SOP Audit

Before you do anything else, you need to know what you're working with. This exercise takes about 90 minutes and will give you a clear picture of your current training content.

Your Assignment:

1. Create a simple spreadsheet with these columns:

2. Set a timer for 60 minutes and start cataloging. Don't get distracted by reading the content. You're just creating an inventory right now. Open your shared drive, your document folders, that random USB drive someone handed you three years ago. List everything.

3. Flag your biggest pain points. After your 60-minute sprint, spend 15 minutes reviewing your list and answering:

- Which procedures do new hires ask about most frequently?
- Which ones have changed but never got updated in the documentation?
- Which ones contradict each other?
- Which ones does nobody follow because they're outdated or impractical?

4. Identify your quick wins. Circle the 3-5 documents that, if converted into interactive training tomorrow, would have the biggest immediate impact on onboarding. These should be:

- Procedures everyone needs to know
- Content that's currently accurate
- Topics that new hires struggle with most

What you'll discover: Most companies find that 20% of their SOPs cover 80% of what new hires actually need to know in their first month. That's where you start. Don't try to convert everything at once. Pick the vital few and nail those first.

Bonus challenge: Share your spreadsheet with three managers and ask them to add a column for "How often do you personally have to explain this?" The procedures they re-explain weekly are your highest-priority conversion targets.

EXERCISE 2: The Module Design Challenge

Pick one of your most important SOPs and turn it into a real training module. This exercise takes about 2 hours and will teach you the process you'll use for everything else.

Your Assignment:

1. Choose one SOP that's 3-10 pages long. Pick something critical but not overly complex for your first attempt:

- Customer service phone script
- Emergency shutdown procedure
- How to process a return
- Your entire HR handbook (too big)
- Complex technical troubleshooting with 50 decision trees (too complex)

2. Break it into 3-5 mini-modules using this structure:

For each mini-module, write down:

Module Title (be specific: "How to Process a Customer Refund" not just "Refunds")

Objective (What will someone be able to DO after this module? Not know, but DO.)

Core Content (List the 3-5 key points or steps)

Real-world Example (When would they actually use this?)

Knowledge Check (Write 2-3 quiz questions)

3. Feed your SOP into an AI tool (ChatGPT, Claude, whatever you use) with this exact prompt:

PROMPT:

"Break this SOP into 5-minute training modules. For each module, provide:

- A clear, specific title
- One actionable learning objective (what the learner will be able to DO)
- The 3-5 most important steps or concepts
- One realistic example of when this would be used on the job

3 quiz questions: one testing recall, one testing process understanding, and one scenario-based

- Make the language conversational and practical for a new hire. Avoid jargon unless it's industry-standard terminology they need to learn."

4. Review what the AI created. Ask yourself:

- Does this make sense to someone who's never done this job?
- Did it miss anything critical?
- Is the language too formal (sounds like a legal document) or too casual (sounds unprofessional)?
- Would someone actually be able to perform this task after completing the module?

Edit it until it sounds like how your best trainer would actually explain this. Read it out loud. If it sounds robotic or weird when you say it, rewrite those parts.

5. Test it on someone. This is the most important step. Grab a colleague who isn't familiar with this specific procedure and have them go through your module. Time them. Can they complete it in 5-10 minutes? After reading it once, can they answer the quiz questions correctly? Most importantly, ask them: "Could you actually do this task now, or would you still need help?"

Get their honest feedback. Write down what confused them. That's where your next revision needs to focus.

What you'll learn: You'll see exactly where your current SOPs are too dense, too vague, or too full of jargon. You'll discover which steps you thought were obvious but actually need more explanation. And you'll have a tested template you can use to transform everything else.

> **PRO TIP:** The first one always takes the longest. Your second module will take half the time because you'll already know what works.

EXERCISE 3: The Quiz Quality Check

Good quizzes separate people who actually learned something from people who just skimmed through. This exercise takes 30 minutes and will dramatically improve your comprehension tracking.

Your Assignment:

1. Look at the quiz questions from Exercise 2. For each question, ask yourself:

2. Rewrite at least one question to make it scenario-based.

Instead of knowledge recall, test application:

Bad: "What is step 3 in the refund process?"
Good: "A customer calls saying they were charged twice for the same order. According to our refund process, what should you do first?"

The second question tests whether they can apply the procedure to a real situation, not just memorize a numbered list.

3. Create one trick question on purpose, then delete it.

Write a question that's intentionally confusing, ambiguous, or tests irrelevant details. For example:

"On what page of the employee handbook is the refund policy located?"

Now delete it. This exercise teaches you what bad quiz questions look like so you can spot them in the future. Trick questions just make people feel stupid without measuring actual comprehension.

4. Have someone take your quiz cold.

Find a colleague who hasn't seen the training module. Have them take your quiz without any preparation.

If they score above 30%, your questions aren't specific enough. They're testing general knowledge instead of content from your module.

If they score below 10% after reading the module, your questions are unfairly tricky or poorly worded.

The sweet spot: Someone with no prior knowledge scores 10-20%, while someone who read the module scores 80%+.

What you'll learn: Writing good quiz questions is harder than it looks. But once you understand the difference between testing comprehension and testing memorization, your entire training system gets exponentially better. You'll start catching yourself when you write questions that don't actually measure understanding.

Challenge question: Can you write a quiz question that has three plausible-sounding answers and one obviously wrong joke answer? Sometimes a little humor helps people relax and engage with the material. Just don't overdo it.

METRICS PACK: Operations Efficiency Tracker

You need to know if this is actually working. Here are the three metrics that matter most when you're transforming your onboarding process with AI.

These aren't vanity metrics. They directly impact your bottom line. Faster onboarding means lower labor costs and quicker time to full productivity. Higher completion rates mean you're not wasting resources on training that people ignore. Better comprehension means fewer mistakes, less rework, and stronger performance from day one.

Track these numbers monthly. Compare them against your pre-AI baseline. When you see onboarding time cut in half and quiz scores climbing into the 85-90% range, you'll know the system is working.

PRO TIP: Don't just track averages. Look at outliers too. If one person took three times longer than everyone else, why? Did they struggle with the content, or did they take breaks between modules? If someone aced every quiz but took forever, they might be overthinking. If someone raced through and bombed the quizzes, they're not engaging with the material.

Module Outline Template + Quiz Prompt

Here are the exact templates we use to turn SOPs into structured training modules. Copy these, adapt them to your content, and you'll have a repeatable system for transforming static documents into interactive learning.

Module Outline Template

Module Title: [Clear, descriptive title that tells learners exactly what they'll learn. Be specific enough that someone could scan a list of 20 modules and know immediately which one they need.]

Objective: What will the learner know or be able to do after completing this module? Be specific and action-oriented.

Vague: "Understand customer refund policy"

Specific: "Process a customer refund request following company policy and document it correctly in the system"

Duration: 5-10 minutes (if it's longer, break it into multiple modules)

Source Document: File name / SOP ID for tracking and updates

Steps:

- [First key step or concept with brief explanation]
- [Second key step or concept with brief explanation]
- [Third key step or concept with brief explanation]

Key Terms: List any technical terms, acronyms, or jargon that need to be defined. Don't assume people know industry terminology.

Common Mistakes: What do people typically get wrong when learning this? Address it proactively.

Real-World Scenario: Give one concrete example of when and how they'd use this on the job.

AI Summary Prompt: "Summarize this module in 3 bullet points for a new hire who needs to understand the essentials quickly."

Quiz Prompt Template

Prompt:

"Create 3 multiple-choice quiz questions based on this module content. Follow these specifications exactly:

Question 1 - Comprehension Test:

Create a question that tests whether the learner understood the core concept. They should be able to answer this if they understood the 'why' behind the procedure, not just the steps.

Question 2 - Process Test:

Create a question that tests whether the learner can recall the correct sequence of steps. Format this as 'What should you do first/next/last?'

Question 3 - Scenario Test:

Create a realistic scenario question that requires applying this knowledge to a situation they might actually encounter on the job. Include enough context that they can't guess the answer without understanding the procedure.

For all questions:

- Provide exactly 4 answer options
- Make 3 options plausible enough that someone who skimmed wouldn't immediately know they're wrong
- Make 1 option clearly correct if someone actually learned the material
- Avoid trick questions, double negatives, or 'all of the above' options
- Keep language clear and concise

Format your response as: Question 1: [question text] A) [option] B) [option] C) [option] D) [option] Correct Answer: [letter] Explanation: [why this is correct]"

When you run this prompt on your module content, the AI generates questions instantly. Review them for accuracy, adjust any wording that feels off, and drop them into your training system. Five minutes of work per module, and you've got built-in accountability that actually measures whether people are learning.

EXERCISE 4: The 30-Day Implementation Sprint

This is your roadmap for going from idea to working system in one month. It's ambitious but totally doable if you commit to the timeline and don't let perfect be the enemy of good.

Week 1: Audit & Prioritize

Monday-Tuesday (4 hours total):
Complete Exercise 1 (The SOP Audit)
Create your master spreadsheet
Identify your top 10 most critical procedures

Wednesday (2 hours):
Meet with 2-3 managers
Get their input on which procedures cause the most questions
Ask: "If you could only train new hires on 5 things, what would they be?"

Thursday (1 hour):
Select your final top 5 SOPs to convert first
These should be: frequently used, currently accurate, and causing the most onboarding friction

Friday (1 hour):
Find one manager willing to pilot this with their next new hire
Set expectations: this is a test, feedback is crucial, we'll adjust as we go

Schedule a 30-minute check-in for Week 4
Week 1 Deliverable: A prioritized list of 5 SOPs ready for conversion and one manager committed to the pilot.

Week 2: Create Content

Monday-Tuesday (6 hours total):

Convert your top 3 SOPs into modules using Exercise 2 process
Don't aim for perfection, aim for functional
Use AI to speed up the initial draft, then edit for your company's voice

Wednesday (3 hours):

Generate quizzes for all modules using the quiz prompt template
Review each question using Exercise 3 quality check
Aim for 3-5 questions per module

Thursday (2 hours):

Convert your remaining 2 SOPs into modules
You should be faster now that you've done three

Friday (2 hours):

Have 2-3 current employees review all modules
Ask specific questions: "Is anything confusing? What's missing? Would you have been able to do your job after reading this on day one?"
Collect their feedback in a simple doc

Week 2 Deliverable: 5 complete training modules with quizzes, reviewed by current employees.

Week 3: Build & Test

Monday-Tuesday (4 hours):

Choose your delivery platform (Notion, ClickUp, shared drive, etc.)
Set up the basic structure: folders, navigation, welcome page
Make it dead simple to find and navigate

Wednesday (3 hours):

Upload all your modules
Add any formatting, images, or visual breaks needed
If using an AI chatbot, integrate it now with instructions trained on your SOPs

Thursday (2 hours):

Go through the entire training flow yourself as if you were a new hire
Time yourself on each module
Take all the quizzes
Note anything confusing, broken, or clunky

Friday (2 hours):

Fix everything you found on Thursday
Have one other person test the full flow
Make final adjustments

Week 3 Deliverable: A fully functional training system tested by at least 2 people.

Week 4: Launch & Measure

Monday (1 hour):

Brief your pilot manager on how the system works
Show them the dashboard where they'll track progress

Answer any questions

Tuesday-Thursday (monitoring):
Have your next new hire go through the system
Check in daily: Are they stuck anywhere? Confused about anything?
Don't intervene unless they're truly blocked, you want to see how it works without handholding

Friday (3 hours):

Collect completion data: How long did each module take? What were quiz scores?
Interview the new hire: What worked? What didn't? Would they recommend any changes?
Interview the manager: How much time did this save them? What would they adjust?

Week 4 Deliverable: Complete data set from one real new hire, including:

- Time to complete each module

- Quiz scores for each module
- Qualitative feedback from both employee and manager
- Comparison to previous onboarding average

What Success Looks Like by Day 30:

You have 5 training modules that work in the real world
You have measurable data showing time saved and comprehension levels
You have a repeatable process for converting more SOPs
You have identified 3-5 specific improvements to make in Version 2
You have one manager who's a believer and will champion this to others

What to do on Day 31:

Make the improvements you identified, then convert your next 5 SOPs. By month two, you should have 10-15 modules. By month three, you've covered 80% of core onboarding. At that point, you're not building anymore, you're just maintaining and optimizing.

THE MANAGER MULTIPLIER

Let me tell you what this really does for your managers.

AI doesn't replace managers. It scales them. Think about what happens in most companies when a new hire starts. A manager or senior employee spends hours walking them through procedures, answering basic questions, and re-explaining things that didn't stick the first time.

Now imagine every new hire had 24/7 access to your best trainer. Someone who never gets tired of answering the same question for the fifteenth time. Someone who can explain procedures at 6 AM or 11 PM, whenever the learner needs it. Someone who adapts to different learning speeds and provides instant feedback on comprehension.

That's what an AI-powered training system does. It takes your institutional knowledge and makes it infinitely replicable.

Your managers stop being walking encyclopedias. They become coaches. Instead of explaining how to fill out a form for the hundredth time, they're mentoring employees on complex problem-solving, helping them develop judgment, and addressing the nuanced situations that actually require human experience.

When one manager figures out a better way to explain a tricky procedure, you update the module once and everyone benefits. When a new policy rolls out, you add it to the system and every employee gets trained on it automatically.

You're not just training people faster. You're building a culture where knowledge is accessible, learning is continuous, and nobody ever has to say, "I didn't know we had a process for that."

THE FUTURE OF TRAINING

In the AI era, knowledge shouldn't sit in folders collecting digital dust.

It should live where people actually learn: on demand, personalized, and measurable.

The companies that figure this out first will have a massive competitive advantage. Not because they have better SOPs. Because they've turned static policies into living systems that make every employee smarter, faster.

Visual Summary: Operations Workflow Map

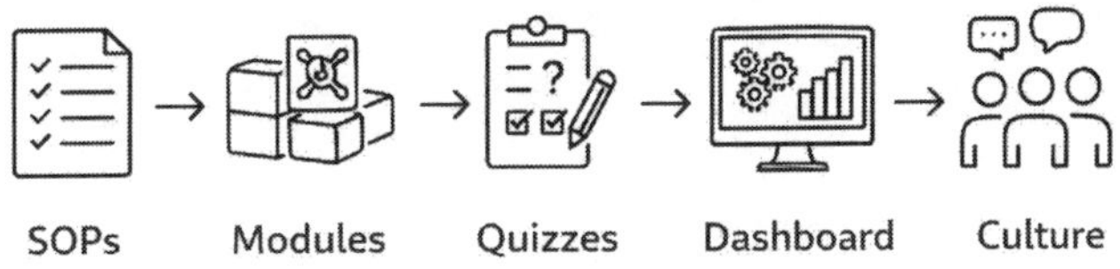

Caption: "AI turns static processes into living learning."

This is the flow. You start with documents nobody reads. AI transforms them into modules people can actually use. Quizzes verify comprehension. Dashboards track progress and highlight gaps. And the result isn't just better training. It's a culture where knowledge is valued, accessible, and continuously improved.

End of Part III Summary

Finance Operations

We've covered how AI transforms your finance operations.

Training Systems

And your training systems.

Next Up? Marketing

If you think AI is good for internal processes, wait until you see how it can help you in the market.

PART IV
Integration and Scale

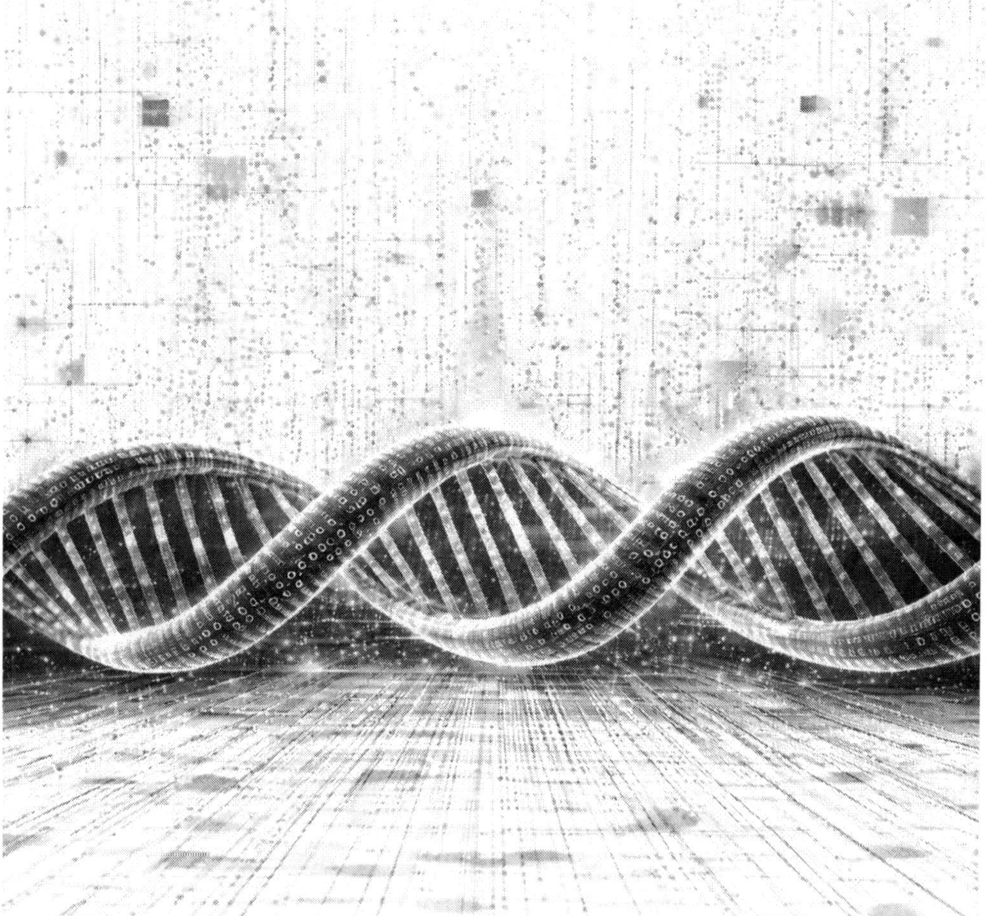

CHAPTER 8

Marketing: Consistent Growth Without More People

> *"You don't need more marketers. You need smarter systems."*
>
> - SUSAN FREW

Marketing used to mean more. More posts, more emails, more meetings, more late nights staring at scheduling grids wondering if anyone actually cared about your Tuesday afternoon LinkedIn post about "workplace culture."

Here's the thing: they didn't.

Not because your message sucked. Because your delivery was inconsistent, your timing was random, and your brand voice sounded like seven different people fighting over one keyboard.

Today, with AI, marketing isn't about volume. It's about consistency and clarity. It's about showing up everywhere your audience is without adding headcount, without burning out your team, and without sounding like a robot wrote your captions at 3 a.m.

AI gives small teams the power of an entire agency. But only if you structure your workflows with focus and guardrails. Because here's what nobody tells you: AI without strategy is just expensive chaos in a prettier package.

Let me show you what actually works.

CASE STUDY: The 3-Hour-a-Week Marketing Engine

I worked with a regional HVAC and plumbing company that was drowning in content creation. They had one marketing coordinator who spent 15 hours a week planning, writing, scheduling, and managing social content across three platforms. She was talented, exhausted, and ready to quit.

Sound familiar?

Her boss kept asking for "more visibility." More posts. More engagement. More proof they were "doing marketing." But here's what was actually happening: the content was inconsistent, the messaging was scattered, and nobody could tell you what the brand voice was supposed to sound like. Every post felt like a coin flip between professional and desperate.

We implemented an AI-driven content system with one human editor. That's it. One person with clear workflows, smart prompts, and permission to trust the system.

Within two months:

- Reduced weekly marketing time from 15 hours to 3 hours.
- Tripled post engagement (yes, tripled. People actually started commenting and sharing).
- Increased website traffic by 47% because content was finally driving clicks instead of crickets.

THE 3C FRAMEWORK FOR AI MARKETING

Before you automate anything, get crystal clear on these three things:

1. Clarity: Know your core messages and audience pain points.
If you can't explain what you do and why it matters in one sentence, your AI can't either. Garbage in, garbage out. Get specific about who you're talking to and what keeps them up at night.

2. Consistency: Automate your publishing rhythm.
Your audience doesn't need daily posts. They need to know you'll show up when you say you will. Weekly. Bi-weekly. Whatever. Pick a cadence and stick to it. AI makes this laughably easy if you let it.

3. Connection: Use AI to tailor messages for each platform, not copy-paste them.
LinkedIn isn't Instagram. Your email list isn't Twitter. AI can rewrite the same core message for different audiences and platforms without sounding like a bot. But you have to ask it to.

When these three align, your brand voice becomes unmissable. And that's when marketing stops feeling like shouting into the void and starts feeling like building relationships at scale.

- Freed leadership to focus on sales conversations instead of arguing about Instagram captions.

The marketing coordinator? She got promoted. Because instead of being buried in busywork, she was finally doing strategic work that moved the business forward.

> *"We stopped chasing trends.*
> *We started building trust, consistently."*
>
> - SUSAN FREW

Don't Let AI Post Directly to Social Media

Here's what gets people in trouble: they think AI can write and post their social content without human oversight.

Wrong.

Social media platforms (LinkedIn, Facebook, Instagram, Twitter) can detect AI-generated content. And when they do? They throttle your reach. Your posts die in the algorithm. You're basically invisible.

I've watched businesses tank their engagement by 60% in two weeks because they automated posting without human editing. The platforms knew. The audience knew. Everyone could tell something was off.

The rule: AI drafts. Humans edit, approve, and post.

AI is brilliant at generating ideas, outlining posts, and creating first drafts. But you need a human to:

Add personality and brand voice

Fact-check and verify accuracy

Adjust tone for current events or audience mood

Make it sound like you, not a content farm

Think of AI as your research assistant and copywriter. Not your social media manager.

Your audience follows you because they trust you. The moment they sense a bot is running your account, that trust evaporates. Don't automate yourself out of relevance.

PLAYBOOK: The AI Content Flywheel

Here's the system that turns one idea into weeks of content without losing your mind or your authenticity.

Step 1: Idea Generation : Prompt Your Brand Voice

Use AI to brainstorm topics, hooks, and outlines based on what your audience actually cares about. Not what you think they care about. What they're asking, searching, and complaining about.

> **PROMPT EXAMPLE:** "Generate 10 short-form content ideas about [your niche] that teach one tip, one mindset shift, or one myth-busting insight in a conversational tone like [brand voice]. Focus on pain points related to [specific challenge]."

> **PRO TIP:** Feed AI examples of your best-performing content first. Show it what "on-brand" actually sounds like. The better your input, the smarter your output.

I've seen teams generate 30 days of content ideas in 20 minutes using this approach. Not filler. Real, valuable, audience-specific ideas they could actually use.

Step 2: Create : Draft Once, Format Many

This is where the magic happens. Produce one long-form piece of content (a blog post, video script, podcast outline, case study, whatever), then use AI to reformat it into shorter content types:

- 3 LinkedIn posts (one educational, one story-driven, one myth-busting)
- 1 newsletter snippet (the hook and one key takeaway)
- 1 Instagram carousel (5-7 slides with one idea per slide)

- 1 30-second video script (tight, punchy, quotable)
- 1 Twitter thread (if that's your jam)

You're not duplicating. You're amplifying. One core message, infinite reach.

Here's what this looked like for that HVAC company: they wrote one 800-word blog post about "Why Your AC Breaks Down in July (And How to Prevent It)." AI turned it into 12 pieces of content across four platforms. Same message. Different formats. Zero burnout.

Step 3: Repurpose : The "AI Echo" Effect

Use AI to rewrite the same core message for different audiences without sounding repetitive or robotic. This is where AI shines, because it can shift tone, language, and framing based on who's reading.

> **PROMPT EXAMPLE:** "Rewrite this paragraph for homeowners aged 40-60 who value reliability over price. Keep it conversational, trustworthy, and avoid technical jargon."

Or:

> "Rewrite this same paragraph for property managers who need fast ROI data and care about tenant satisfaction metrics."

Same facts. Different frame. Authentic variety.

This is how you create content that feels personal without writing 47 versions of the same post by hand. AI does the heavy lifting. You do the quality control.

Step 4: Distribute : The Autopilot Layer

Use tools like Zapier, Buffer, Later, or Hootsuite to schedule content generated from AI templates. Set your posting cadences once a week. Let AI fill the gaps.

Here's the workflow:

- Monday morning: Review AI-generated content for the week.
- 30 minutes of edits: Add your voice, adjust tone, approve final versions.
- Load into scheduler: Set it and forget it.
- Track engagement throughout the week: Note what lands, what doesn't.

The HVAC company did this every Monday at 9 a.m. Three hours. Done. Content locked and loaded for the entire week. No scrambling. No last-minute "we need a post today" panic.

And here's the kicker: their engagement went up because the content was finally consistent. People started expecting to hear from them. That's when trust builds.

Step 5: Measure : Feedback, Not Vanity Metrics

Track content that drives clicks, conversions, or saves. Not just likes. Likes don't pay bills.

AI analytics tools can summarize engagement trends and suggest next steps weekly. You don't need a data science degree. You need a dashboard that tells you what's working and what's wasting your time.

Ask your AI:

"Summarize last week's content performance. Which formats drove the most engagement? Which topics resonated most? What should we double down on next week?"

You'll get a readable summary in 60 seconds. Use it. Adjust. Repeat.

LINKEDIN MASTERY: How to Actually Get Found (And Why TLDR Matters)

LinkedIn isn't social media. It's a search engine for professionals. And if you're not optimizing for search, you're invisible.

Here's what most people get wrong: they treat LinkedIn like Facebook. Short posts. Quick takes. Random thoughts. That stuff gets buried.

LinkedIn's algorithm rewards long-form content. Deep, valuable posts that keep people on the platform. Posts that spark conversations. Posts that demonstrate expertise.

The TLDR Strategy

When you write a long-form LinkedIn post (500-1,200 words), always include a TLDR (Too Long; Didn't Read) at the top or bottom.

Why? Because you're serving two audiences:

The skimmers: People scrolling fast who need the headline version.

The deep readers: People who want the full story and actionable insights.

Example Structure:

TLDR: [One-sentence summary of your main point]

[Full post with story, data, insights]

Want the short version? Scroll up to the TLDR.

Or flip it:

[Hook that grabs attention in first 2 lines]

[Full post]

TLDR: [Bullet points of key takeaways]

This does three things:

- Increases engagement: People comment on both the summary and the details.
- Boosts dwell time: LinkedIn tracks how long people spend reading your post. Longer = better visibility.
- Builds authority: Long-form content signals expertise. You're not just posting. You're teaching.

How to Use AI for LinkedIn Content (The Right Way)

Step 1: Draft your post in AI

Give it your core message, your story, your data points.

> **PROMPT EXAMPLE:** "Write a 700-word LinkedIn post about [topic] that opens with a relatable story, transitions to 3 key lessons, and closes with one actionable step readers can take today. Use short paragraphs. Keep the tone conversational and authoritative. Include a TLDR at the end."

Step 2: Humanize it

Read it out loud. Does it sound like you? Add your personality. Change robotic phrases. Insert your opinion.

Step 3: Add engagement hooks

Ask a question in the first line
Use line breaks generously (white space = readability)
End with a call to action: "What's worked for you?" or "Thoughts?"

Step 4: Post manually

Never use AI to auto-post. LinkedIn knows. And it will kill your reach.

VIDEO: The Non-Negotiable Marketing Asset

Let's be brutally honest: if you're not doing video, you're losing.

Video drives more engagement than any other content type. Period. LinkedIn prioritizes video. Instagram prioritizes video. Facebook prioritizes video. YouTube is the second-largest search engine in the world.

Your audience wants to see your face, hear your voice, and trust that you're a real human who knows what you're talking about.

And here's the thing: it doesn't have to be perfect. It has to be consistent.

Why Video Matters More Than Ever

Trust accelerator: People buy from people they trust. Video builds trust faster than text ever will.

Algorithm love: Platforms push video to more people because it keeps users on the platform longer.

Accessibility: Video + captions reaches more people than text alone.

Repurposing goldmine: One 10-minute video can become 20 pieces of content.

The AI Video Workflow

Step 1: Script it with AI

PROMPT EXAMPLE: "Write a 90-second video script about [topic] that opens with a problem, offers one solution, and ends with a clear next step. Keep sentences short and conversational. Write for spoken word, not reading."

AI will give you a script. Read it out loud. Adjust it until it sounds like you talking, not you reading.

Step 2: Record it (imperfectly)

Use your phone. Natural light. Decent audio. Done.
You don't need a studio. You need clarity and consistency.

Step 3: Caption it with AI

Upload your video to tools like Descript, Otter.ai, or Riverside.fm. AI will transcribe it. Edit the captions for accuracy. Export.

Step 4: Repurpose it everywhere

Post the full video on YouTube
Post a 60-second clip on LinkedIn
Post a 30-second teaser on Instagram
Pull 5 quote graphics from the transcript
Turn the transcript into a blog post
One video. Ten assets.

Video Types That Work

Behind-the-scenes: Show your process, your team, your workspace.

Customer stories: Let your customers tell their transformation story.

Tips and how-tos: Teach one thing in 90 seconds.

Myth-busting: "Everyone thinks X, but here's the truth."

Weekly updates: Consistency builds trust. Pick a day. Show up every week.

PRO TIP: Don't overthink it. The videos that perform best are often the ones recorded on phones with zero editing. Authenticity beats production value every single time.

THE FUTURE OF SEARCH: AEO and GEO Explained

Google is changing. Fast.

Traditional SEO (Search Engine Optimization) is still important. But it's not enough anymore.

Now we have:

AEO (Answer Engine Optimization): Optimizing for AI-powered answer engines

GEO (Generative Engine Optimization): Optimizing for AI-generated search results

Translation: when someone asks ChatGPT, Claude, Gemini, or Perplexity a question about your industry, you want your business to be the answer.

Here's how to do it.

PLAYBOOK: The AEO/GEO Formula

What Is AEO (Answer Engine Optimization)?

AEO is about structuring your content so AI tools can easily find, understand, and cite you as an authoritative source.

When someone asks an AI, "What's the best way to prevent HVAC breakdowns in summer?" you want AI to reference your content.

The AEO Formula (Step-by-Step)

Step 1: Identify Your Core Questions

What questions does your audience ask most often?

Use:

- Customer support emails
- Sales call recordings
- Google "People Also Ask" section

- Reddit threads in your industry
- Quora searches
- Make a list of 20-30 questions your customers repeatedly ask.
- Step 2: Create Direct, Authoritative Answers
- For each question, create content that:
- Answers the question in the first 2-3 sentences
- Provides clear, step-by-step guidance
- Uses simple, conversational language
- Includes data, examples, or case studies

> **AI PROMPT EXAMPLE:** "Write a 500-word blog post that answers the question: [insert question]. Start with a direct answer in the first paragraph. Use subheadings for readability. Include one real-world example. Write at an 8th-grade reading level."

Step 3: Structure for AI Readability

AI tools scan for patterns. Make it easy:

- Use clear headings: H2s and H3s that match common questions
- Add FAQ sections: Literally write "Frequently Asked Questions" with direct answers
- Format with lists: Bullet points and numbered lists are AI-friendly
- Include definitions: Define industry terms clearly
- Add schema markup: Use FAQ or How-To schema on your website (your web developer can help)

Step 4: Publish Everywhere

Don't just put it on your blog. Distribute the same answers across:

- LinkedIn posts
- YouTube video descriptions
- Medium articles
- Your email newsletter

- Industry forums

The more places your answer exists, the more likely AI will cite you.

What Is GEO (Generative Engine Optimization)?

GEO is optimizing for AI-generated search summaries. When someone Googles a question, Google now uses AI to generate a comprehensive answer at the top of the search results (called SGE: Search Generative Experience).

You want your content to be in that AI-generated answer.

The GEO Formula (Step-by-Step)

Step 1: Focus on E-E-A-T

Google's AI prioritizes content that demonstrates:

- Experience: You've actually done this, not just read about it
- Expertise: You know what you're talking about
- Authoritativeness: Others recognize you as credible
- Trustworthiness: Your information is accurate and reliable

How to show E-E-A-T:

- Include author bios with credentials
- Link to authoritative sources
- Use real data and case studies
- Get backlinks from credible sites in your industry
- Step 2: Write for Featured Snippets
- Google's AI pulls from featured snippets. Optimize for them:
- For "What is" questions: Write a 40-60 word definition in the first paragraph
- For "How to" questions: Use numbered steps with clear headings
- For "Best" questions: Create comparison lists or tables
- For "Why" questions: Provide a clear explanation followed by supporting points

Example:

How to Prevent HVAC Breakdowns in Summer

- Step 1: Schedule a pre-season inspection in April or May
- Step 2: Replace air filters every 30-60 days during peak use
- Step 3: Clear debris from outdoor units weekly
- Step 4: Monitor thermostat settings and avoid drastic temperature changes
- Step 5: Keep vents unobstructed throughout your home

AI loves this format. Clean. Clear. Actionable.

Step 3: Create Comprehensive Content

Google's AI favors depth. Write 1,500-2,500 word articles that cover a topic thoroughly.

Include:

- Introduction with direct answer
- Background and context
- Step-by-step guidance
- Common mistakes to avoid
- Tools and resources
- Real examples or case studies
- FAQ section
- Conclusion with next steps

> **AI PROMPT FOR COMPREHENSIVE CONTENT:** "Create a comprehensive 2,000-word guide on [topic] that covers: definition, why it matters, step-by-step process, common mistakes, best tools, real-world example, and FAQ. Structure it for maximum readability with clear headings and short paragraphs."

Step 4: Optimize for Semantic Search

AI doesn't just look for keywords. It understands context and intent.

Instead of stuffing keywords, write naturally and include:

Synonyms: If your keyword is "HVAC repair," also use "air conditioning fix," "heating system maintenance"

Related topics: Cover adjacent questions people might ask

Entities: Mention specific brands, locations, or industry terms

Context: Explain why something matters, not just what it is

Step 5: Update Regularly

AI prioritizes fresh content. Update your top-performing articles every 3-6 months with:

- New data or statistics
- Recent case studies
- Updated best practices
- Current tool recommendations
- Set a calendar reminder. Treat content like inventory. Keep it fresh.

METRIC PACK: Marketing ROI Dashboard

Evaluate by effort-to-impact ratio. If a format takes too long for minimal reach, kill it fast. I don't care if everyone says you "need to be on TikTok." If it's not moving your business forward, it's a vanity project. Cut it.

ARTIFACT: AI Campaign Template

Campaign Name:
(e.g., Fall HVAC Tune-Up Awareness)

Objective:
(e.g., Drive 20% more service calls in October)

Core Message:
(What's the one story you're telling?)

Key Audiences:
(e.g., Homeowners 40-60, Property Managers)

Formats:

- Blog (AEO/GEO optimized)
- Video (3 versions: long, medium, short)
- Email
- Social Carousel
- LinkedIn Post (with TLDR)

AI Prompts Used:
(List the prompts for idea generation, tone matching, and repurposing.)

Publishing Cadence:
(Monday – LinkedIn long-form / Wednesday – Email + Video / Friday – Instagram)

SEO/AEO Keywords:
(Primary keyword, 3-5 related terms, target questions)

Measurement Cadence:
(Weekly review via AI analytics summary.)

End of Campaign Review:
☐ Keep ☐ Iterate ☐ Retire

Use this template every single time. Consistency compounds.

STEP-BY-STEP: Your First AEO/GEO-Optimized Content Piece

Let me walk you through creating one piece of content that's optimized for both traditional search and AI-generated answers.

Week 1: Research and Planning

Day 1-2: Question Mining

Open Google and type your main service/product + "how to"

Note the "People Also Ask" questions

Check Reddit, Quora, and industry forums for common questions

Review your customer support emails from the last 90 days

Create a list of 10 questions you can answer better than anyone

Day 3: Competitor Analysis

Google your top 3 questions

Read the top 5 results for each

Note what's missing or unclear in those answers

Identify gaps you can fill with better information

Day 4: Content Planning

Choose ONE question for your first piece. Pick the one where:

You have the strongest expertise

There's high search volume (use Ubersuggest or AnswerThePublic for free)

Existing answers are weak or incomplete

AI PROMPT: "Create a detailed content outline for an article answering [your question]. Include: hook, main sections, key points to cover, real-world examples needed, and FAQ section. Make it comprehensive but readable."

Week 2: Creation

Day 1: Write the Core Content

Use your outline. Write 1,500-2,000 words.

Start with a direct answer in the first paragraph. Then dive deeper with:

Why this matters

Step-by-step guidance

Common mistakes

Pro tips from experience

> **AI PROMPT:** "Expand this outline into a full article. Keep paragraphs short (2-3 sentences). Use conversational language. Include transition sentences between sections. Write at an 8th-grade reading level."

Then humanize it. Add your voice. Insert personal examples. Make it sound like you.

Day 2: Add Structure

Insert clear H2 and H3 headings

Add a table of contents at the top

Create a TLDR box

Add an FAQ section at the end (5-7 questions)

Include relevant images or diagrams

Day 3: Optimize for AI

Add schema markup (FAQ schema is easiest)

Include internal links to related content

Add external links to 2-3 authoritative sources

Write a meta description (150-160 characters) that includes your main keyword

Create an SEO-friendly URL slug

> **AI PROMPT FOR META DESCRIPTION:** "Write 3 variations of a meta description for this article. Each should be 150-160 characters, include the keyword [your keyword], and entice clicks."

Week 3: Distribution

Day 1: Publish to Your Website

Upload to your blog. Make sure it's mobile-friendly.

Day 2-3: Create Spin-Off Content

Use AI to repurpose your article into:

One 10-minute YouTube video (using the article as a script)

One LinkedIn long-form post (with TLDR)

Three quote graphics for Instagram

One email to your list

> **PROMPT:** "Turn this 2,000-word article into a 10-minute video script. Keep the structure but make it conversational for spoken word. Include natural pauses and emphasis points."

Day 4-5: Share Strategically

Post to:

Your website blog

LinkedIn (tag relevant people and companies)

Medium (republish with canonical link back to your site)

Relevant LinkedIn groups or Facebook groups

Industry forums (where helpful, not spammy)

Week 4: Measure and Iterate

Day 1: Check Initial Performance

How many page views in the first week?

What's the average time on page?

Any comments or shares?

Where is it ranking for your target keyword?

Day 2: Ask AI for Analysis

> **PROMPT:** "Based on this performance data [paste your data], what should I change or improve about this content? What's working? What's not?"

Day 3-4: Make Updates

Based on feedback:

Clarify confusing sections

Add more examples

Expand weak areas

Update with fresh data

Day 5: Plan Your Next Piece

Pick the next question from your list. Repeat the process.

By the end of 90 days, you'll have 10-12 deeply optimized pieces of content that position you as the go-to expert in your space. And AI tools will start citing you as a source.

> **YOUR BRAND VOICE IS YOUR MOAT**
>
> AI can amplify your voice, not replace it.
>
> Create a Brand Voice Prompt Library. Your words, your tone, examples of what "on-brand" feels like. Feed this to AI every time you generate content.
>
> Example:
>
> "Our brand voice is conversational, confident, and practical. We sound like a smart friend who's done the research and cut through the BS. We use short sentences. We avoid jargon. We value clarity over cleverness. We tell stories. We don't preach."
>
> Then add 3-5 examples of content that nails your voice. Show AI what success looks like.
>
> Sound like yourself, not like AI. That's your competitive edge.

The Marketing Dividend

When AI automates the repetition, your team can focus on resonance. Crafting stories that build community, not just clicks. Creating moments that matter, not just posts that check boxes.

That's the dividend. Time. Energy. Creativity. The stuff that actually grows businesses.

Marketing Workflow Map

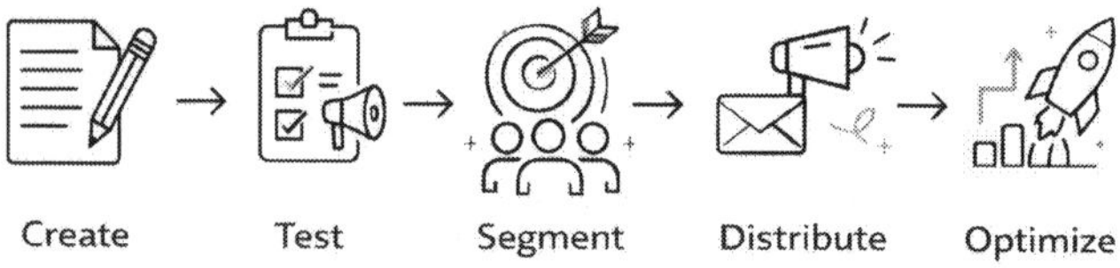

Your 30-Day Marketing Transformation Checklist

Week 1: Foundation

☐ Create your Brand Voice Prompt Library

☐ Identify your top 10 customer questions

☐ Set up your content calendar (pick your posting days)

☐ Choose your primary content format (blog, video, or LinkedIn)

Week 2: Creation

☐ Create your first AEO/GEO-optimized article

☐ Record your first video (even if it's 60 seconds)

☐ Write 3 LinkedIn posts using the TLDR method

☐ Build your AI prompt library for content repurposing

Week 3: Distribution

☐ Publish your article with proper optimization

☐ Repurpose it into 5+ content pieces

☐ Set up your scheduling system (Buffer, Later, etc.)

☐ Share across 3+ platforms with platform-specific edits

Week 4: Measurement

☐ Review engagement metrics

☐ Ask AI to analyze what's working

☐ Update content based on feedback

☐ Plan next month's topics and schedule

Repeat monthly. Consistency is everything.

Closing Insight

AI doesn't make you louder. It makes you clearer.

When marketing becomes a system instead of a scramble, your business finally breathes. You show up consistently, confidently, and creatively without burnout. Without begging your team to "just post something" on Friday afternoons. Without wondering if anyone's even listening.

Video makes you trustworthy. Long-form content makes you findable. AEO and GEO make you the answer AI provides when someone asks about your industry.

And that's the ultimate marketing advantage: consistency that compounds.

One idea. One system. Infinite reach.

Now go build your flywheel. And for the love of all things holy, don't let AI post directly to social media. Edit first. Always.

The First Three Workflows

CHAPTER 9

Integrate (Connect the Dots Before You Add More Tools)

Here's what nobody tells you about AI implementation: The problem isn't that you don't have enough tools. It's that the ones you already have aren't talking to each other.

I see this pattern play out constantly with the executives and entrepreneurs I work with. They've read the articles, attended the webinars, signed up for the free trials. Their tech stack looks impressive on paper. But behind the scenes? It's chaos. Data living in seventeen different places. Three people doing the same task in three different systems. AI tools that promised to save time actually creating more work because nothing connects.

Integration isn't sexy. It doesn't come with flashy demos or clever marketing campaigns. But it's the quiet power move that separates companies actually succeeding with AI from those just accumulating expensive subscriptions.

Think of integration like this: You can have the best ingredients in your kitchen, but if your oven doesn't work with your pans and your knife doesn't match your cutting board, you're going to spend more time wrestling with equipment than actually cooking. AI works the same way. One good tool connected to your data is worth ten that aren't.

The companies winning with AI right now? They're not the ones with the most tools. They're the ones who figured out how to make fewer tools work harder.

The $1,200 Subscription Problem

Let me tell you about a conversation I had last fall with the COO of a mid-sized professional services firm. Smart guy, forward-thinking, genuinely invested in making AI work for his team. He'd been early to adopt AI tools and felt pretty good about his company's tech-forward approach.

Until I asked him one simple question: "How many AI subscriptions are you currently paying for?"

He paused. "Maybe... six? Eight?"

We spent the next hour doing an audit. Turns out, they were paying for seventeen separate AI tools. Seventeen. And here's the kicker: at least six of them were doing essentially the same thing. They had three different AI writing assistants (one for marketing, one someone in sales had signed up for, and one the legal team swore they needed). They had four separate tools for meeting transcription because different teams preferred different platforms. They were using two competing project management systems with AI features because nobody wanted to migrate their data.

The monthly cost? Just over $3,400. But the real cost wasn't the money. It was the fragmentation. Marketing couldn't see what sales was working on. The project management data didn't sync. Meeting notes from one tool couldn't be searched by people using another tool. They'd created an expensive AI archipelago where every team was on their own island.

After we consolidated into four connected core systems (their existing CRM with AI features activated, a unified workspace tool, one transcription service that fed into everything else, and a properly integrated analytics dashboard), they cut their monthly AI spend by $1,200. More importantly, they could actually find things. Data flowed between systems. People stopped wasting time reformatting information to move it from one tool to another.

Six months later, that COO told me the integration work had delivered more value than any single AI tool they'd implemented. Because now those tools could actually work together.

Quote Box: "AI tools are like gym memberships. Everyone signs up, few actually use them. And somehow you end up paying for three when you only need one."

The lesson? More tools don't equal more productivity. They equal more complexity. And complexity is the enemy of adoption.

The "Less Is More" Rule

Visual: Simplified tech stack diagram showing 3 integrated tools vs. 12 disconnected ones

I've developed a simple filter for evaluating whether a new AI tool is worth adding to your stack. Every tool you're considering should pass this three-part test:

Does it eliminate an existing manual step?

If the tool doesn't remove something tedious from someone's plate, it's just adding to their workload. I once watched a company add an AI scheduling assistant that still required a human to confirm every meeting manually. That's not automation. That's just extra steps with fancier packaging.

Does it connect seamlessly with your existing data sources?

A tool that requires manual data entry or constant copy-pasting isn't an upgrade. It's a bottleneck with good branding. If it can't pull from and push to your existing systems, you're building another data silo.

Can you measure clear ROI within 30 days?

Notice I didn't say "potential ROI" or "estimated value over time." I mean actual, measurable results in the first month. Time saved. Tasks completed faster. Errors reduced. Money directly saved or earned. If you can't point to concrete numbers in 30 days, you're dealing with innovation theater, not business transformation.

If a tool can't check all three boxes, it's not a solution. It's innovation clutter.

I know this sounds harsh. The AI marketplace is full of impressive-looking tools with compelling demos. But your job isn't to collect AI tools. Your job is to run a business. And every tool that doesn't integrate creates friction that slows everything down.

Here's what integration thinking looks like in practice: Instead of adding a new AI customer service chatbot that lives on a separate platform, you evaluate whether your existing CRM can handle chatbot functionality. Instead of signing up for another AI writing tool because it has one cool feature, you check if your current system can be configured to do something similar. Instead of letting every department choose their own AI solutions, you create a core stack that everyone builds from.

This doesn't mean you never add new tools. It means you're strategic about what you add and ruthless about what you eliminate. The best tech stacks aren't the biggest ones. They're the tightest ones.

Playbook: Build Your Smart Stack

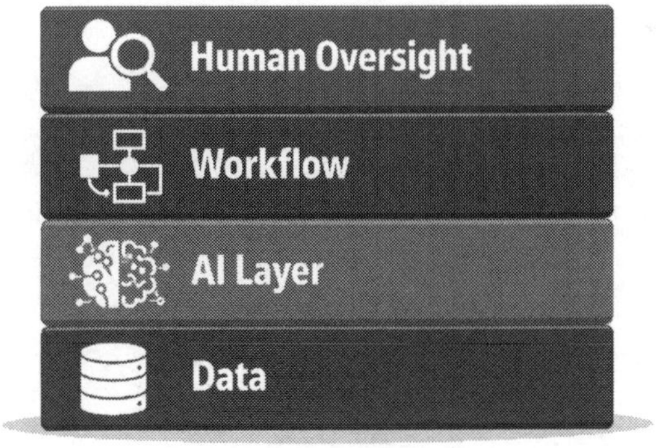

Most companies approach their tech stack backwards. They pick tools they like and then try to force their data and workflows to fit. That's like buying furniture first and then building a house around it. It creates weird spaces, wasted resources, and a lot of unnecessary frustration.

Instead, start with your foundation and build up deliberately. Here's the five-step process I walk clients through when we're building (or rebuilding) their AI infrastructure:

Step 1: Map Your Data Ecosystem

Before you can integrate anything, you need to know what you're working with. And I mean everything. Not just the official systems, but the actual places people store and access information daily.

Grab a whiteboard or open a fresh doc and start listing:

- Your CRM (and any satellite customer databases people have created)
- Cloud storage (Google Drive, Dropbox, OneDrive, that shared folder someone set up three years ago)
- Email systems and how they're organized
- Project management platforms
- Financial software
- Communication tools (Slack, Teams, whatever channels people actually use)
- Spreadsheets and databases (the official ones and the ones people maintain on their desktops)
- Industry-specific software
- Any tools that house customer data, contracts, or historical records

Don't skip the informal systems. Those rogue spreadsheets your sales director maintains? That's part of your data ecosystem. The shared Google Doc where the operations team tracks everything because they don't trust the official system? That counts too.

This step usually reveals uncomfortable truths. You'll discover data living in places you didn't know existed. You'll find the same information stored in three different formats across five different systems. You'll realize that half your team is working off last month's data because nothing syncs.

Good. Now you know what you're dealing with.

Step 2: Choose a Core Platform

Once you've mapped everything, you need to pick your "AI hub." This is the single source of truth where your most critical data lives and where your AI tools will connect.

For some companies, this is their CRM. For others, it's a workspace tool like Notion or Airtable. Some businesses build their hub in their project management system. The specific platform matters less than the criteria:

It needs to handle your most frequently accessed data

Your team actually uses it (not what they're supposed to use, what they actually open daily)

It has solid integration capabilities via API or native connections

It can scale with your business without requiring a complete rebuild in two years

This is not the time to chase the newest, shiniest platform. Choose something stable, well-supported, and proven. You're building your foundation here. Boring and reliable beats cutting-edge and unstable every single time.

One warning: Resist the urge to choose different hubs for different departments. I know marketing wants Notion, sales loves their CRM, and operations is married to their project tracker. But multiple hubs defeat the whole purpose of integration. Pick one. Make it work for everyone. The short-term complaints are worth the long-term efficiency.

Step 3: Connect Workflows via API or Zapier

Now comes the part where everything starts talking to each other. This is where automation stops being theoretical and starts being functional.

You're looking for the repetitive data transfers that happen manually right now. Things like:

- New customer signs up → data entered in CRM → welcome email sent → task created for onboarding

- Invoice generated in QuickBooks → notification sent to team → payment tracked in spreadsheet
- Meeting notes taken → action items extracted → tasks assigned → follow-up scheduled
- Form submitted on website → lead captured → routed to sales → added to pipeline

Each of these workflows currently requires human hands at multiple steps. With proper integration, they can flow automatically.

Start with one high-impact workflow. Map out every step. Then figure out how to automate the transitions between steps. For example:

Manual process:

New invoice created in QuickBooks → someone downloads it → someone emails it to the client → someone updates the tracking spreadsheet → someone adds a calendar reminder for follow-up

Integrated process:

New invoice created in QuickBooks → automatically triggers AI summary generation → summary posted in Slack channel → dashboard updates automatically → follow-up task auto-assigned based on client tier

You don't need to become a developer to make this work. Tools like Zapier, Make, or native integrations handle most of the heavy lifting. The trick is thinking through the workflow first, then finding the technical solution second.

Step 4: Add Human Checkpoints

Here's where a lot of companies mess up integration: they automate everything and remove all human oversight. Then something goes wrong, and nobody notices until it's a disaster.

Automation doesn't mean abdication. It means strategically removing tedious manual steps while keeping humans in the loop at critical decision points.

Think of it like this: automate the repetitive, preserve the thoughtful. Let AI and automation handle data transfer, formatting, initial drafts, routine calculations, and status updates. Keep humans involved in final approvals, client-facing communications, strategic decisions, and anything involving judgment calls.

For every automated workflow you create, ask: "Where does a human need to review, approve, or intervene?" Then build that checkpoint into the system. Maybe it's an approval step before anything gets sent to a client. Maybe it's a daily summary that shows what automation did overnight. Maybe it's a flag system that alerts people when something falls outside normal parameters.

These checkpoints aren't bottlenecks. They're guardrails. They let automation run fast while keeping you in control.

Step 5: Review Monthly

Integration isn't a one-time project. It's an ongoing practice.

Set a standing monthly "Tech Stack Review" meeting. Thirty minutes. Same time every month. Non-negotiable. Bring the same core group: someone from operations, someone technical, and someone who actually uses the tools daily.

Review three things:

What's working? Which integrations are saving time? Where is data flowing smoothly? Which tools are earning their subscription cost?

What's broken? Where are the friction points? What's requiring workarounds? Where are people reverting to manual processes because the automation isn't working?

What's redundant? Are multiple tools doing the same job? Has something become obsolete because we added a new feature elsewhere? Can we consolidate?

Track your findings. Make decisions. Cancel subscriptions. Fix broken integrations. Celebrate wins.

This monthly rhythm keeps your tech stack lean and functional. It prevents the slow accumulation of forgotten subscriptions and abandoned tools. And it ensures that as your business evolves, your integration strategy evolves with it.

Integration Performance Dashboard

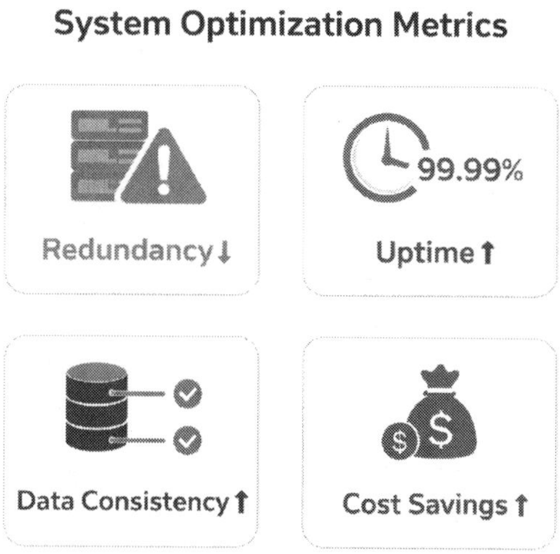

You can't manage what you don't measure. Here are the four metrics that actually tell you whether your integration strategy is working:

Here's how to track these:

Tool Redundancy: Count your AI and automation tools at the start. Set a goal to cut that number in half within 90 days. Track it monthly. If you're not reducing the count, you're not integrating. You're just adding.

Data Sync Accuracy: Pick five critical data points (customer information, financial figures, project status, whatever matters most to your business). Spot-check them weekly across all systems where they should appear. Calculate what percentage match. If you're below 95%, your integrations have serious gaps.

Monthly Cost Savings: Total up all your AI tool subscriptions right now. Set a baseline. Then track your monthly spend. A good integration strategy should reduce costs by 25% or more within the first quarter, not by cutting capability but by eliminating redundancy and negotiating better pricing for the tools you keep.

Integration Uptime: Track how often your key integrations fail or require manual intervention. Log it. If you're spending more time fixing integrations than you're saving through automation, something's wrong with your setup.

Pro Tip: Don't measure tools by how many features they have. Measure them by how much friction they remove. The tool with 47 AI features that requires constant troubleshooting is worth less than the simple tool with three features that works flawlessly every single time.

AI Integration Map Template

Visual: Network diagram with boxes labeled Tool, Function, Owner, ROI

Here's a template I use with every client to map their current state and plan their ideal state. Fill this out honestly. Don't list what you wish you were using or what the vendor demos showed. List what's actually happening in your business right now.

Current State Integration Map

Fill in your version. Be brutal. The "Actual ROI" column is where most companies discover uncomfortable truths. That expensive tool you signed up for in January? If nobody's using it, the ROI is zero regardless of its potential.

Target State Integration Map (90 Days Out)

Notice what's different? Fewer tools. More connections. Clearer owners. Measurable targets.

This isn't aspirational. This is your 90-day roadmap. Every tool on the target state map should have a specific integration plan and a person responsible for making it happen.

The Compound Effect of Connection

Here's what integration really gives you: leverage.

One AI tool working in isolation might save someone an hour a week. That same tool connected to your data and integrated into your workflows? It can save your entire team ten hours a week. It can eliminate entire categories of errors. It can surface insights you'd never find manually.

Integration turns individual tools into a system. And systems compound.

When your meeting notes automatically feed into your project tracker, which updates your dashboard, which triggers relevant alerts to the right people, you're not just saving time on each step. You're eliminating the entire category of work called "keeping everyone updated." Information flows without human intervention. Context gets preserved. Nothing falls through the cracks.

That's the compound effect. Small integrations create big impacts when they stack together.

The companies I work with who've truly mastered AI integration describe a shift in how work feels. Less time spent hunting for information. Fewer meetings required to align everyone. Reduced anxiety about whether things are getting done because the system shows you.

That only happens when you stop collecting tools and start connecting them.

Scale comes from flow, not volume.

Chapter Summary

Integration isn't about having the fanciest AI tools. It's about making the tools you have work together seamlessly. Cut the redundancy, connect your core systems, add strategic human checkpoints, and review regularly. The companies winning with AI aren't the ones with the most subscriptions. They're the ones who figured out how to make fewer tools work harder through smart integration. One well-connected tool beats ten that operate in isolation. Every time.

CHAPTER 10

Train & Transform : From Early Adopters to AI Culture

Your AI pilots worked. Your executives are happy. Your early adopters are winning. And absolutely none of that matters if the rest of your company is still working like it's 2019.

Here's the truth most consultants won't tell you: your AI strategy doesn't fail because of technology. It fails because you treated adoption like a one-time training event instead of a cultural transformation. You brought in some expert for a two-hour workshop, handed out a deck full of buzzwords, and wondered why nothing changed.

Culture doesn't shift because someone showed up with a PowerPoint. It shifts when people see their colleagues winning, when they realize AI makes their jobs easier instead of harder, and when they actually get the support they need to figure this thing out.

This chapter is about turning your three AI champions into 300 believers. It's about creating the kind of environment where people experiment freely, share openly, and actually want to learn this stuff. And it's about building systems that make AI adoption inevitable instead of optional.

Let's get to work.

CASE STUDY: From 3 Champions to 300 Believers

A mid-sized regional healthcare organization started small. Three people. Three departments. Three separate AI pilots running quietly in the background while everyone else pretended not to notice.

Finance used AI to process invoices faster. Customer service deployed a chatbot that actually worked. Operations built a scheduling assistant that saved the team eight hours a week.

Nothing revolutionary. Nothing flashy. Just three people proving that AI could make work easier.

Then something interesting happened.

Other employees started asking questions. Not hostile questions like "Is this going to replace me?" but curious ones like "Wait, how did you do that?" and "Could this work for my team?"

The company made a choice. Instead of keeping those three champions siloed, they turned them into teachers. They created a shared AI library where anyone could grab prompts, workflows, and templates. They scheduled short weekly sessions where champions showed real examples from their actual work, not theoretical use cases from some consultant's deck.

Six months later? Three hundred employees trained. A company-wide AI knowledge base that people actually used. And measurable results that showed up in their quarterly reports, not just in some aspirational slide deck.

Finance cut invoice processing time from 4 days to 6 hours. Customer service reduced response time by 60%. Operations saved 416 hours per quarter, which translated to real budget they could reinvest. The overall productivity increase? A documented 28% lift that executives could actually point to in board meetings.

The breakthrough wasn't the technology. It was the culture shift from "those people are doing that AI thing" to "we're an organization that uses AI."

PLAYBOOK: The Train-and-Transform Blueprint

Here's your step-by-step system for scaling AI adoption from a handful of champions to company-wide culture. This isn't theory. This is what actually works.

Step 1: Identify AI Champions

Every department needs one curious, trusted team member to lead adoption. Not the most senior person. Not necessarily the most tech-savvy. The person everyone already goes to when they need help figuring something out.

Look for people who:

Experiment naturally instead of waiting for perfect instructions
Share knowledge without being asked
Stay calm when things break
Get excited about efficiency gains

You're not looking for volunteers who want a resume bullet. You're looking for people who genuinely enjoy solving problems and helping their teammates win.

Give them a title if your culture requires it. "AI Integration Lead" or "Digital Transformation Champion" or whatever makes sense in your organization. More importantly, give them time, resources, and permission to experiment without needing three levels of approval.

THE 4 STAGES OF AI CULTURE

Every organization moves through four predictable stages when building AI culture. You can't skip stages, but you can accelerate them.

Stage 1: Awareness
People see AI's potential but haven't touched it yet. They're watching your pilots, reading the headlines, wondering if this matters for their work. Your job here isn't convincing. It's showing.

Stage 2: Adoption
They start using it daily, even if it's just one or two simple tasks. They're still figuring it out, still making mistakes, still not entirely sure this is worth the effort. Your job here is support, not pressure.

Stage 3: Advocacy
They share results and teach others. They're the ones saying "you should try this" in team meetings. They're building templates and volunteering to help colleagues. Your job here is amplification, not micromanagement.

Stage 4: Autonomy
AI becomes second nature across teams. People create their own workflows, share discoveries, and improve processes without needing permission or hand-holding. Your job here is staying out of the way.

Most companies get stuck between Awareness and Adoption because they treat training like a checkbox instead of a journey. They blast everyone with information, expect instant mastery, and wonder why adoption stalls.

Your goal isn't control. It's confidence.

Step 2: Create an AI Playbook Library

Store prompts, SOPs, and dashboards in one searchable location that people can actually find and use. Not buried in SharePoint. Not scattered across email threads. One place.

Your library should include:

- Proven prompts with context (when to use them, what they accomplish)
- Step-by-step workflows for common tasks
- Real examples from your company, not generic templates
- Video walkthroughs (2-3 minutes max)
- Common mistakes and how to fix them

Organize by department and function, not by AI tool. Someone in Finance shouldn't need to know which platform does what. They should type "invoice processing" and find exactly what they need.

Example naming convention:

Finance → Invoice AI Workflow → SOP v2.3

Version numbers matter. People need to know they're looking at the current process, not something someone created two years ago and forgot to update.

Step 3: Build Mini Training Moments

Fifteen-minute sessions weekly beat two-hour quarterly workshops every single time. Microlearning beats marathon sessions because people retain what they practice, not what they sit through.

Structure your mini-sessions like this:

5 minutes: Here's the problem this solves
7 minutes: Watch me do it live (not a polished demo, actual work)
3 minutes: Questions and next steps
Record everything. Not everyone can attend live, and people rewatch when they're ready to try it themselves.

Rotate presenters. Let your champions teach. Let early adopters share their wins and their mistakes. The best training comes from peers showing real work, not executives reading prepared remarks about "digital transformation."

Step 4: Reward Early Adopters

Celebrate small wins publicly. Incentivize team members who share success stories, create useful prompts, or help colleagues get unstuck.

Recognition matters more than money, though money doesn't hurt. Try:

Shoutouts in company meetings

"AI Win of the Week" email series
Bonus points in performance reviews
First pick at new tools or pilot programs
Small gift cards or team lunches

Make it clear that sharing knowledge is valued, not hoarding it. The person who creates a prompt that saves the team 50 hours isn't competing with colleagues. They're making everyone more effective.

Step 5: Track Organizational Lift

Measure output per hour, task turnaround time, and training engagement. Not because you want to micromanage, but because you need proof that this matters.

Track what actually indicates progress:

How many people are using AI tools daily (not just signed up)
How often people access your playbook library
How much time teams save on repetitive tasks
How many new workflows get created and shared
How training completion correlates with productivity gains

Don't measure just to measure. Measure to identify what's working, what's stuck, and where you need to focus next.

METRIC PACK: Culture Growth Indicators

These aren't vanity metrics. They're diagnostic tools. If adoption rate is high but productivity lift is flat, your tools aren't solving real problems. If training completion is low but library usage is climbing, people prefer self-service learning over formal sessions.

Watch the trends. Adjust your approach. Stay focused on outcomes, not activity.

ARTIFACT: The AI Culture Charter

A culture charter isn't a poster on the wall. It's a living document that defines how your organization commits to learning, sharing, and improving with AI.

AI Culture Charter Template

Your charter should be specific enough to drive action but flexible enough to evolve. Review it quarterly. Update it when you learn something new. Make sure people actually know it exists.

> **PROMPT ADD-ON:** "Summarize this week's AI success stories in three sentences for our internal newsletter."

Use AI to help you celebrate AI adoption. It's meta, and it works.

THE 70/20/10 RULE

Here's how adults actually learn new skills:

70% of learning happens by doing. Not watching tutorials. Not reading documentation. Actually using the tool, making mistakes, figuring it out.

20% happens by sharing with peers. Asking questions in Slack. Watching a colleague solve a problem. Grabbing coffee with someone who's already figured this out.

10% comes from formal training. Workshops, courses, certification programs. Important, but not sufficient.

Most companies flip this ratio. They dump everything into formal training and wonder why nothing sticks.

Design your AI adoption strategy around the 70/20/10 balance. Give people time to experiment. Create space for peer learning. Use formal training as a foundation, not the entire structure.

Callout Box: The Human Advantage

The future isn't humans versus AI. It's humans plus AI.

The companies that win are the ones that train both, together. Not AI replacing people. Not people resisting AI. People becoming more capable, more creative, more valuable because they know how to use these tools effectively.

Your competitive advantage isn't the technology. Every company can buy the same platforms. Your advantage is how well your people use them, how quickly they adapt, and how naturally they integrate AI into their actual work.

Build that culture, and you won't just survive the AI transformation. You'll lead it.

VISUAL SUMMARY: The Scale System

The Scale System

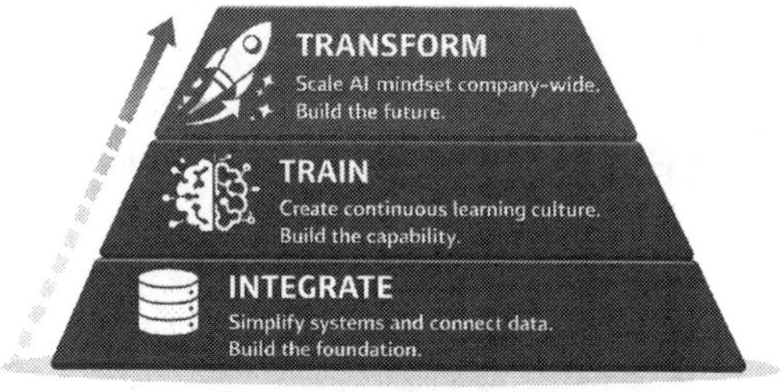

- ✓ Each layer widens reach and deepens adoption.
- ✓ You can't skip levels. You can't rush the process.
- ✓ But you can build deliberately, measuring progress, celebrating wins, and creating the kind of organization that learns faster than the competition.

End Of Part Iv Summary

You've made it through the hardest part: moving from proof of concept to organizational transformation. Here's what you've built:

You started with skeptics. You converted some champions. You built systems that make adoption easier. And now you're creating a culture where AI isn't something people resist or tolerate. It's something they expect, use, and improve.

That's not just implementing technology. That's transforming how your company thinks, works, and competes.

Building AI culture isn't optional anymore. It's not a nice-to-have or a future initiative. It's the difference between companies that adapt and companies that become cautionary tales. You've done the work. Now make it stick.

PART V

The Future: Agents, Ethics, and Autonomy

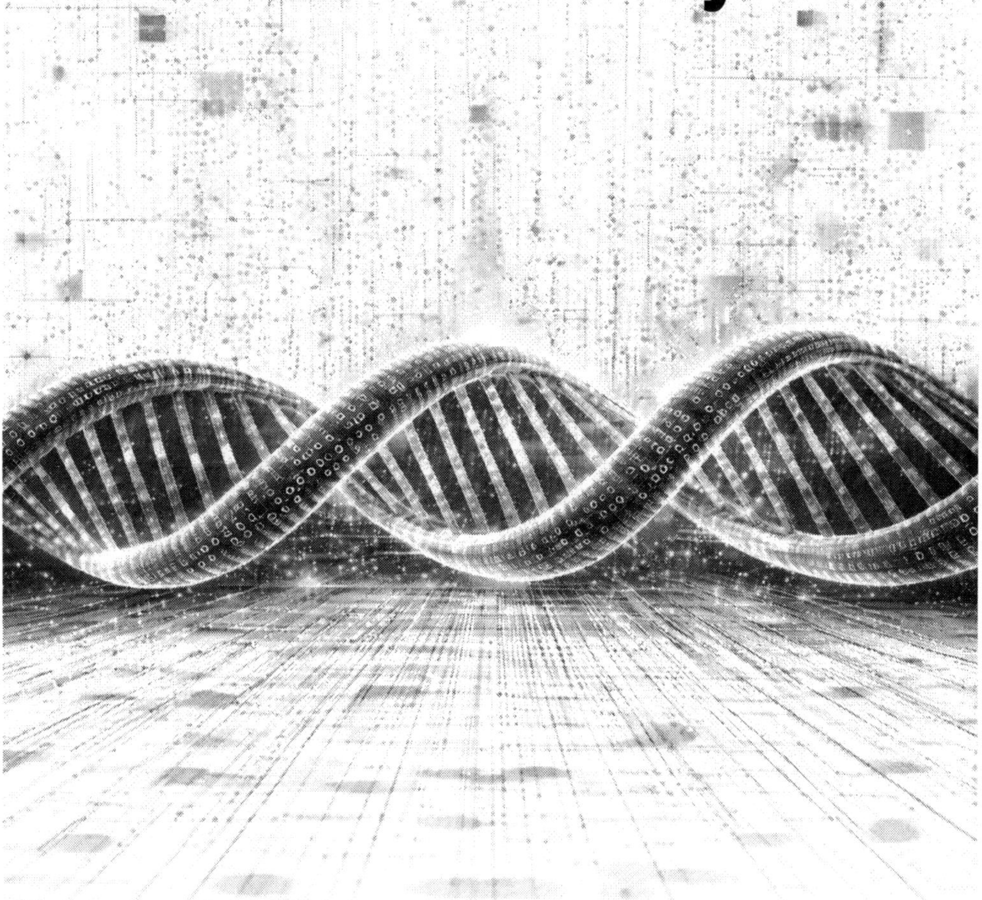

CHAPTER 11

Agents: The Rise of the Digital Team

Let me tell you what keeps most executives up at night: it's not that AI will replace their team. It's that they won't know how to use it fast enough to stay competitive.

Here's the truth nobody's saying out loud: the next phase of AI isn't just about better prompts or smarter chatbots. It's about agents : autonomous systems that don't just answer questions or generate ideas. They take action.

These AI agents can schedule meetings, analyze data patterns, summarize reports, respond to communications, and manage entire workflows on your behalf. And here's what makes them different from the automation tools you're already using: they learn. They adapt. They remember context. They collaborate.

Think of them less like software and more like the most reliable junior team member you've ever had : one who never gets tired, never takes a sick day, and doesn't need three follow-up emails to finish a task.

But here's what matters most: AI agents don't replace your employees. They extend them.

They handle the repetitive, soul-crushing tasks that drain your team's energy and creativity. The stuff that has to get done but doesn't move the needle. The work that keeps your best people from doing their actual best work.

That's the promise. Now let me show you what it looks like in practice.

CASE STUDY: One Manager, Five AI Assistants

A mid-sized professional services company came to me with a problem every growing business faces: their department heads were drowning in administrative work. These weren't junior managers. These were experienced leaders who should have been focused on strategy, client relationships, and team development. Instead, they were buried in email, scheduling, data entry, and report generation.

Sound familiar?

We tested something radical: an internal "AI Team" model. Each department head got one specialized AI agent trained on their specific workflows, communication style, and decision-making patterns.

The Finance Agent handled transaction summaries, anomaly detection, and monthly report preparation.

The HR Agent managed interview scheduling, candidate communications, and onboarding checklists.

The Marketing Agent drafted social posts, analyzed campaign performance, and maintained the content calendar.

The Ops Agent tracked project milestones, flagged bottlenecks, and automated status updates.

The Admin Agent coordinated meetings, managed inbox triage, and handled routine correspondence.

We ran this for 90 days. Here's what happened:

3,000 hours of repetitive work automated. That's not a typo. When we tracked time saved across five department heads, the number was staggering. Tasks that used to take 15 minutes now took 30 seconds.

25% faster response times to customers. When your team isn't buried in busywork, they can actually respond to the people who matter. Customer inquiries that used to sit for hours got answered within minutes.

Consistent tone and process across departments. This was the unexpected win. Every agent was trained on company standards, so communications sounded professional and on-brand whether they came from Finance or Marketing.

But here's my favorite result, the one that doesn't show up in a spreadsheet: every single department head said they felt like they could think again.

They weren't just getting tasks done faster. They were getting their attention back. Their creativity back. Their ability to actually lead instead of just manage.

"My AI team doesn't take vacations : but it makes sure my humans can." : Susan Frew

That's the real value of AI agents. Not just efficiency. Freedom.

AGENTS VS. AUTOMATION

Let's clear up some confusion. People hear "AI agents" and think it's just fancy automation. It's not.

Automation is pre-programmed and rigid. It follows a fixed process. If A happens, do B. Every single time. No exceptions. No adaptation. It's great for repetitive tasks that never change.

Agents are adaptive and context-aware. They learn from patterns, remember previous interactions, and adjust their approach based on the situation. They're problem solvers, not just task executors.

Think of it this way: automation is a vending machine. You push the same button, you get the same snack. An AI agent is more like a personal assistant who knows you prefer your coffee black on Mondays but with cream on Fridays because Fridays are rough.

Agents learn. They remember. They collaborate. They're the next level of workflow scaling, and they're already here.

PLAYBOOK: Building Your AI Team

Ready to build your own AI team? Good. Here's how to do it without losing your mind or your budget.

Step 1: Identify Repetitive Knowledge Work

Start by making a list of tasks your team does over and over again. Not physical tasks (robots aren't assembling your products yet), but knowledge work. Things like:

- Responding to routine emails
- Scheduling meetings across time zones
- Summarizing reports or meeting notes
- Generating weekly status updates

- Tracking project milestones
- Creating first drafts of standard documents
- Pulling data from multiple systems into one report

These are agent-ready tasks. Write them down. You probably have more than you think.

Step 2: Assign "Roles" Not "Tasks"

Here's where most people get it wrong. They think about AI agents as task machines: "Make this AI do X." That's old-school automation thinking.

Instead, define agent roles the same way you'd define job descriptions for humans.

Don't say: "I need an AI to send calendar invites."

Say: "I need a Scheduling Coordinator Agent who manages my calendar, handles meeting requests, finds optimal times across time zones, sends invites with agendas, and follows up with participants."

See the difference? You're giving the agent a job, not just a task. Jobs have context. Jobs require judgment. Jobs evolve.

Step 3: Give Context, Not Just Prompts

AI agents perform best when they understand the bigger picture. They need to know:

Your company's goals. Is speed more important than perfection? Is brand consistency non-negotiable? What are you actually trying to accomplish?

Your SOPs and style guides. How does your company communicate? What's your tone? What are your formatting standards? Give your agents the same onboarding documents you'd give a new hire.

Decision rules. When should the agent take action autonomously? When should it ask for input? When should it escalate to a human? Be specific.

For example: "If a customer inquiry mentions billing issues or amounts over $500, flag it for immediate human review. Otherwise, draft a response and send it."

The more context you provide upfront, the less babysitting you'll do later.

Step 4: Start in Sandbox Mode

Do not, under any circumstances, turn your AI agents loose on live customer communications or critical workflows on Day One. That's how nightmares happen.

Instead, run them in shadow mode for at least 30 days. Let them draft emails without sending them. Let them create reports that you review before sharing. Let them schedule meetings that you confirm before they go out.

Compare their output to what your humans would have done. Track accuracy. Identify patterns in errors. Adjust their instructions.

Think of this as your agent's probationary period. You're not just testing the technology. You're building trust.

Step 5: Promote to Production

Once you're confident in your agents' performance, integrate them into live systems. But even then, build in human review checkpoints for anything high-stakes.

For routine, low-risk tasks? Let them run autonomously. For anything involving money, legal issues, or sensitive communications? Keep a human in the loop.

You're not choosing between human control and AI autonomy. You're designing the right balance for each workflow.

METRIC PACK: Agent Productivity Dashboard

Here's how you know if your AI agents are actually working or just creating expensive busywork:

If your agents aren't hitting these targets within 60-90 days, either your workflows aren't agent-ready yet or you need to adjust their instructions.

Don't keep feeding a broken system. Fix it or cut it.

ARTIFACT: AI Agent Role Card Template

Every AI agent in your organization should have a role card. Think of it like a job description mixed with an instruction manual. Here's the template:

Agent Name: Finance Analyst Agent

Purpose: Summarize daily transactions, detect spending anomalies, and prepare weekly financial reports.

Inputs: Access to accounting software, expense dashboards, and company financial policies.

Triggers:

Weekly reports generated every Friday at 5 PM
Monthly close procedures on the last business day
Keyword alerts for transactions flagged as "urgent" or "review needed"
Escalation Rules:
Send alerts to CFO for any single transaction over $10,000
Flag unusual spending patterns (>20% variance from baseline)
Escalate vendor payment issues to accounting team lead
Performance Metrics:
Accuracy above 95% (spot-checked monthly)
Report completion time under 10 minutes
Zero missed deadlines
Review Schedule: Monthly performance review with Finance Director

Create one of these for every agent you deploy. It keeps everyone aligned on what the agent should (and shouldn't) be doing. It also makes it easy to train new team members on how to work with your AI team.

The Digital Team Advantage

Here's what most people get wrong about AI agents: they think the goal is to save time.

Time savings matter. But that's not the real advantage.

The real advantage is attention.

Your best people are juggling way too much. They're context-switching between strategic projects and administrative garbage. They're spending 60% of their day on work that doesn't require their expertise, experience, or creativity.

AI agents don't just save time. They extend attention. They make sure the right human is focused on the right thing, at the right time.

Your Marketing Director shouldn't be scheduling social posts. She should be crafting campaign strategy.

Your Operations Manager shouldn't be updating spreadsheets. He should be identifying process improvements.

Your Sales Lead shouldn't be writing follow-up emails. She should be closing deals.

AI agents handle the noise so your humans can focus on the signal.

That's not about replacing people. That's about freeing them to do what they were actually hired to do.

And if you're thinking, "But my team likes being busy" or "We're not big enough for this yet," let me be blunt: your competitors are already doing this. The question isn't whether to build a digital team. The question is whether you'll do it before you're forced to catch up.

CHAPTER 12

Ethics & Autonomy: Building Trust in the Machine Age

The more powerful AI becomes, the more critical trust becomes.

Ethics isn't a compliance checklist you tick off once and forget. It's a leadership muscle you build, strengthen, and exercise every single day. How you train your people, communicate your standards, and monitor AI use will determine whether your organization thrives in the machine age or fractures under the weight of broken trust.

Here's the truth: Your customers don't fear AI because it's intelligent. They fear it because they don't understand how you're using it or what you're doing with their data. That fear costs you money, loyalty, and market share.

So let's fix that.

CASE STUDY: The Transparency Dividend

Two similar companies rolled out customer-facing chat systems within weeks of each other. Both systems delivered fast, accurate responses. Both saved time and money.

The difference? One company disclosed when AI was being used. The other kept it hidden, hoping customers wouldn't notice.

The result: The transparent company saw a 22% increase in customer trust scores and higher internal adoption rates. The secretive one faced backlash, complaints, and employee skepticism that took months to repair.

Transparency isn't a liability. It's a competitive advantage.

PLAYBOOK: The Ethics Guardrail Framework

Building ethical AI use isn't complicated. It just requires intention and consistency. Here's your five-step framework for putting guardrails in place before something goes wrong.

Step 1: Declare Your AI Policy

Publish a simple, public statement that sets expectations. No jargon. No legalese. Just clarity.

Example: "We use AI to assist our team, not replace them. Human oversight remains the final decision-maker in all customer interactions."

Post it on your website. Share it in onboarding. Make it impossible to miss.

Step 2: Protect Data Like Gold

Use private or enterprise versions of AI tools whenever possible. Never feed customer data, proprietary information, or sensitive details into free public models.

If you wouldn't post it on a billboard, don't put it in ChatGPT.

THE ETHICAL TRIANGLE

Every AI decision should balance three forces:

1. People: Who is affected by this decision? What impact does it have on their work, privacy, or trust?

2. Process: How is data handled? Who reviews outputs? What safeguards exist to prevent misuse or error?

3. Transparency: Can you explain the "why" behind the AI's recommendation in language anyone can understand?

If you can't answer all three clearly, you're not ready to deploy.

Step 3: Audit Outputs Regularly

AI drift is real. Models change. Prompts degrade. Bias creeps in.
Set a monthly review process to evaluate agent behavior, output quality, and data integrity. Catch problems before your customers do.

Step 4: Create a Feedback Channel

Give employees and customers an easy, safe way to report concerns about AI use. Make it anonymous if needed. Act on what you learn.

Trust grows when people see you listening.

Step 5: Train for Judgment

Teach your team not just how to use AI, but when not to. Not every task needs automation. Not every output is worth trusting.

Good judgment isn't automatic. It's trained.

METRIC PACK: Trust and Compliance Tracker

Track these metrics the same way you track revenue. What you measure, you manage.

ARTIFACT: AI Ethics Toolkit (Condensed Field Guide)

This toolkit gives your team everything they need to operate ethically without slowing down.

1. Policy Template

"Our AI systems are designed to augment human work, not replace it. We commit to transparency, accountability, and rigorous data protection in everything we build."

Adapt this to your brand voice. Post it publicly. Live it internally.

2. Audit Checklist

Before approving any AI-generated output, ask:

- Is this factually accurate?
- Is any sensitive data exposed?
- Would I sign my name to this?

If the answer to any of these is "no" or "I'm not sure," don't ship it.

3. Disclosure Prompt

Insert this line at the end of every AI-assisted report, email, or customer communication:

"This content was created with the assistance of AI and verified by [Name]."

Simple. Honest. Trust-building.

4. Review Cadence

Quarterly: Leadership reviews AI use and outcomes.
Annually: Full ethics refresh with updated training.
Immediately: Flag and investigate any anomaly or complaint.
Consistency builds credibility.

TRUST IS YOUR BRAND

In the AI era, your greatest differentiator isn't speed, volume, or cost savings.

It's trust:the confidence that what you create is both smart and safe.

When customers trust you, they stay. When employees trust you, they innovate. When regulators trust you, they leave you alone.

Trust isn't a nice-to-have. It's everything.

Closing Reflection: From Tool to Transformation

Visual: Human and AI hand meeting at a beam of light

AI is no longer a sidekick or an experiment. It's woven into how we think, how we lead, and how we create value in the world.

But as we scale into an age of intelligent agents and autonomous systems, the defining factor won't be what AI can do. It will be what we choose to do with it.

The companies that win won't be the ones who moved fastest. They'll be the ones who moved wisest:building systems their customers trust, their employees respect, and their industries admire.

> *"AI doesn't replace human ingenuity.*
> *It expands it:one smart, ethical choice at a time."*
>
> - SUSAN FREW

VISUAL SUMMARY: The Future Framework

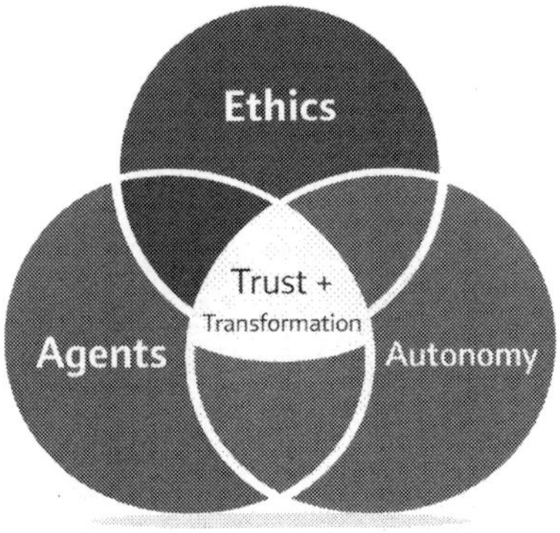

The future isn't faster. It's wiser.

End of Part V Summary

History Assessment

[ ] We regularly discuss our origin stories and pivotal moments
[ ] We understand why our key processes are still in place
[ ] We maintain traditions that work and eliminate those that don't

Questions to Answer:

- What old decision is still shaping how you operate today?
- Which founding story drives current behavior?
- What legacy process needs to be challenged?

Habits Assessment

[ ] We can name our top three unwritten rules
[ ] We know which routines help and which waste time
[ ] We actively experiment with new ways of working

Questions to Answer:

- What habit would you change first? Why?
- What's the gap between official policy and actual practice?
- Which informal workflows are actually running the business?

Tech Stack Assessment

[ ] Our tools work together and integrate seamlessly
[ ] We've documented all systems and assigned clear ownership
[ ] We evaluate new tools based on impact and security, not features

Questions to Answer:

- Which tool do you use most? Which tool do you use least?
- Where are the integration gaps?
- What shadow IT exists in your organization?

People Assessment

[ ] Everyone knows their role and responsibilities clearly
[ ] Teams have diverse skills and perspectives
[ ] We invest consistently in ongoing development

Questions to Answer:

Who carries your culture? Why them?
Where are the unofficial power networks?
What skills are missing from your team?

A2. 30-Day AI Pilot Charter Template

Pilot Name: ______________________________

Owner: ______________________________ (Name & Role)

Start Date: //20___ End Date: //20___

Goal: (One measurable outcome)

Success Metric:

Baseline Data: (Current state before pilot)

Tools/Models Used:

[ ] ChatGPT Team

[ ] Claude Projects

[ ] Other: ____________________________

Resources Required:

Time: _____ hours/week

Budget: $_____

Integrations needed: ____________________________

Training required: ____________________________

Weekly Review Schedule:

Week 1: //20___

Week 2: //20___

Week 3: //20___

Week 4: //20___

End-of-Pilot Decision:

[ ] Kill (abandon this approach)

[ ] Keep (scale to broader team)

[ ] Iterate (refine and test again)

Lessons Learned:

A3. Finance Workflow Automation Checklist

Phase 1: Invoice Management

[ ] Set up automated invoice intake (via email or upload)
[ ] Configure OCR tool for data extraction
[ ] Establish vendor categorization rules
[ ] Define approval thresholds
[ ] Create exception handling workflow
[ ] Set up human review checkpoints

Phase 2: Reconciliation

[ ] Connect accounting system to AI tools
[ ] Define matching parameters
[ ] Set variance thresholds for flagging
[ ] Create daily digest format
[ ] Establish escalation procedures

Phase 3: Reporting

[ ] Build AI-powered summary templates
[ ] Schedule automated report generation
[ ] Define KPIs to track
[ ] Set up dashboard access
[ ] Create distribution list

Success Metrics:

Cost per document processed: Target ↓50%
Invoice cycle time: Target <3 days
Exception rate: Target <10%
Team satisfaction: Target ↑30%

A4. Operations Training Module Template

Module Title: ____________________________

Learning Objective: (What will learners be able to DO?)

Duration: 5-10 minutes

Source Document: ____________________________

Key Steps:

Common Mistakes to Avoid:

Real-World Scenario:

Quiz Questions:

Q1 (Comprehension): ____________________________

A) ____________________________

B) ____________________________

C) ____________________________

D) ____________________________

Correct Answer: ___

Q2 (Process): ____________________________

A) ____________________________

B) ____________________________

C) ____________________________

D) ____________________________

Correct Answer: ___

Q3 (Scenario): ____________________________

A) ____________________________

B) ____________________________

C) ____________________________

D) ____________________________

Correct Answer: ___

A5. Marketing AI Campaign Template

Campaign Name: ________________________

Objective: (Specific, measurable goal)

Core Message: (One story you're telling)

Target Audiences:

Content Formats:

[ ] Blog post (AEO/GEO optimized)
[ ] Video (long-form 5-10 min)
[ ] Video (short-form 30-60 sec)
[ ] Email newsletter
[ ] LinkedIn long-form post (with TLDR)
[ ] Social carousel
[ ] Infographic
[ ] Podcast/audio

AI Prompts Used:

Publishing Cadence:

Monday: ________________________

Wednesday: ________________________

Friday: ________________________

SEO/AEO Keywords:

Primary: ________________________

Secondary: ________________________

Question targets: ________________________

Measurement Schedule:

Weekly review via AI analytics

Monthly performance summary

Quarterly campaign assessment

End-of-Campaign Review:

[ ] Keep (successful, repeat approach)
[ ] Iterate (partial success, needs adjustment)
[ ] Retire (didn't meet objectives)

A6. Integration Performance Scorecard

Monthly Tech Stack Review

Date: //20___

Tool Redundancy Check:

Current number of AI tools: _____

Tools performing similar functions: _____

Consolidation opportunities: _____

Target: ↓50% reduction in redundant tools

Data Sync Accuracy:

Critical data points checked: _____

Consistency rate across systems: _____%

Target: 95%+ consistency

Monthly Cost Analysis:

Total AI/tech spending: $_____

Cost vs. baseline: _____%

Target: ↓25% reduction

Integration Uptime:

Key integrations monitored: _____

Uptime percentage: _____%

Manual interventions required: _____

Target: 99%+ uptime

Action Items:

A7. AI Culture Charter

Our AI Culture Commitments

Experiment Freely

Commitment: We encourage safe testing and learning, even when experiments fail

Owner: Department Leads

Review: Ongoing

Share Openly

Commitment: We publish learnings in our internal hub and celebrate contributions

Owner: AI Champions

Review: Weekly

Train Consistently

Commitment: We include AI refreshers in every onboarding and team meeting

Owner: HR & Training

Review: Monthly

Measure Transparently

Commitment: We track metrics and celebrate growth, not perfection

Owner: Executive Team

Review: Quarterly

Signature/Date: ______________________ //20___

A8. AI Agent Role Card Template

Agent Name: ______________________

Purpose:

Primary Functions:

Data Access:

[ ] Accounting software
[ ] CRM
[ ] Email
[ ] Project management
[ ] Other: ______________________

Trigger Events:

Escalation Rules: When to alert a human:

Performance Metrics:

Accuracy target: ____%

Completion time: _____ minutes

Error rate: <____%

Review Schedule:

Last Updated: //20___

A9. AI Ethics Toolkit

AI Ethics Policy Statement

"Our organization uses AI to augment human work, not replace it. We commit to transparency, accountability, and rigorous data protection in all AI applications. Human oversight remains the final authority in customer-facing decisions."

Pre-Deployment Audit Checklist

Before deploying any AI-generated content:

[ ] Is this factually accurate?
[ ] Is any sensitive data exposed?
[ ] Does this align with our brand voice?
[ ] Would I personally sign my name to this?
[ ] Has a qualified human reviewed this output?
[ ] Are we disclosing AI use appropriately?

Disclosure Language

For internal documents:
"This content was created with AI assistance and verified by [Name], [Title]."

For customer communications:
"We use AI tools to help our team serve you faster. All responses are reviewed by our staff to ensure accuracy."

Review Cadence
Weekly: Spot-check AI outputs for quality

Monthly: Review agent performance metrics

Quarterly: Leadership ethics and compliance review

Annually: Full AI policy refresh and team training

B. Prompt Libraries

B1. Finance Prompts

Invoice Processing:

"Extract the following from this invoice: vendor name, invoice number, date, line items with descriptions and amounts, total, payment terms. Format as structured data."

Variance Analysis:

"Analyze this month's spending data against budget. Identify categories with >10% variance and provide a 3-sentence explanation for each significant variance."

Financial Summary:

"Summarize this financial data for executive review. Include: key trends, anomalies requiring attention, and recommended actions. Keep under 200 words."

B2. Operations Prompts

SOP Conversion:

"Break this SOP into 5-minute training modules. For each module provide: specific title, one actionable objective, 3-5 key steps, one realistic example, and 3 quiz questions testing comprehension."

Process Improvement:

"Analyze this workflow description. Identify steps that are: repetitive, time-consuming, prone to error. Suggest which steps could be automated with current AI tools."

B3. Marketing Prompts

Content Calendar:

"Generate 30 days of social content ideas for [industry/audience]. Focus on [key topics]. Include: post topic, format (video/image/text), key message, and suggested CTA. Tone: [brand voice]."

Long-Form LinkedIn:

"Write a 700-word LinkedIn post about [topic] that opens with a relatable story, transitions to 3 key lessons, and closes with one actionable step. Use short paragraphs. Include a TLDR at the end."

AEO/GEO Optimization:

"Create a comprehensive 2,000-word guide on [topic] that covers: definition, why it matters, step-by-step process, common mistakes, best tools, real-world example, and FAQ section. Structure for maximum readability."

B4. Customer Service Prompts

Response Drafting:

"Draft a response to this customer inquiry: [paste inquiry]. Tone should be helpful and professional. Address their concern directly in the first paragraph, provide solution, and include relevant next steps."

Issue Categorization:

"Categorize this support ticket: [paste content]. Assign priority (High/Medium/Low), department (Sales/Support/Technical), and suggest initial response type."

C. Tool & Resource Directory

AI Platforms

Conversational AI:
ChatGPT (OpenAI) - Team: $30/user/month
Claude (Anthropic) - Team/Projects available
Google Gemini - Enterprise tiers available
Microsoft Copilot - Integrated with Microsoft 365
Finance & Accounting:
Dext (Receipt Bank) - Invoice processing, ~$40/month
Hubdoc (Xero) - Document management
Bill.com - Accounts payable automation
Botkeeper - AI bookkeeping service
Expensify - Expense management
Workflow Automation:
Zapier - Integration platform, starts free
Make (Integromat) - Advanced workflow automation
n8n - Open-source alternative
Document & Knowledge Management:
Notion - Workspace with AI features
Airtable - Database with automation
Trainual - SOP and training platform
ClickUp - Project management with AI
Marketing & Content:
Descript - Video editing and transcription
Canva - Design platform with AI features
Buffer - Social media scheduling
Later - Instagram-focused scheduling
Hootsuite - Multi-platform management
Communication & CRM:
Intercom - Customer messaging with AI
Drift - Conversational marketing
HubSpot - CRM with AI features
Salesforce Einstein - AI for Salesforce

Data & Analytics:
Tableau - Data visualization
Power BI - Microsoft analytics platform
Looker - Google Cloud analytics

D. 90-Day Implementation Roadmap

Month 1: Foundation (Discovery & Design)

Week 1: Audit Current State
Map existing workflows (Finance, Ops, Marketing)
Identify tool redundancies
Baseline key metrics
Select pilot team (3-5 champions)
Week 2: Prioritize & Plan
Choose 3 Monday-ready workflows
Design safety guardrails
Create vendor evaluation checklist
Draft AI acceptable use policy
Week 3: Build Foundation
Set up core platform/hub
Establish data governance rules
Create basic prompt library
Schedule weekly check-ins
Week 4: Launch Pilots
Deploy first workflow in sandbox mode
Track baseline vs. AI-assisted performance
Document early wins and failures
Prepare for Month 2 expansion

Month 2: Pilot & Measure

Week 5: Expand Testing
Add second workflow to pilot
Train pilot team on documentation
Create feedback loops

Share early results with leadership
Week 6: Iterate & Refine
Adjust based on Week 1-5 learnings
Build integration connections
Develop internal playbook structure
Plan Wave 2 training
Week 7: Train Wave 2
Department leads and managers
Share pilot successes and lessons
Distribute templates and prompts
Set expectations for rollout
Week 8: Quality Assurance
Spot-check AI outputs
Review compliance and safety
Measure KPIs against targets
Prepare Month 3 scaling plan

Month 3: Scale & Optimize

Week 9: Broader Rollout
Train frontline teams
Deploy successful workflows company-wide
Establish ongoing governance
Create feedback channels
Week 10: Integration Focus
Connect AI tools to core systems
Eliminate redundant subscriptions
Optimize data flows
Document integration architecture
Week 11: Culture Building
Launch internal AI knowledge hub
Recognize and reward champions
Share success stories publicly
Address resistance and concerns
Week 12: Review & Plan
Measure overall productivity lift

Calculate ROI and cost savings
Kill/Keep/Iterate decision on each workflow
Plan next quarter's expansion

E. Glossary of AI Terms

Agent: An AI system that can take autonomous action based on goals and context, not just respond to prompts.

AEO (Answer Engine Optimization): Structuring content so AI tools can easily find, understand, and cite it as authoritative.

Bias: Systematic errors in AI output resulting from skewed training data or flawed algorithms.

Fine-tuning: Customizing a pre-trained AI model with specific data to improve performance on targeted tasks.

GEO (Generative Engine Optimization): Optimizing content for AI-generated search summaries (like Google's SGE).

Hallucination: When AI generates plausible-sounding but factually incorrect information.

Human in the Loop: Workflow design where AI performs tasks but humans review and approve outputs.

Integration: Connecting AI tools to existing systems and data sources for seamless workflow.

LLM (Large Language Model): AI trained on massive text datasets to understand and generate human language.

OCR (Optical Character Recognition): Technology that converts images of text into machine-readable text.

Pilot: Small-scale test of an AI workflow before full deployment.

Prompt: Instructions given to an AI system to generate a desired output.

Schema Markup: Code added to websites helping search engines understand content structure.

Shadow IT: Technology tools used within an organization without official approval or oversight.

Token: Unit of text processed by AI (roughly 0.75 words in English).

NOTES

The "Business DNA" framework draws on organizational theory from Strategy+Business, "The DNA of Innovation," December 2003, which compared corporate DNA to biological DNA with four key bases affecting organizational behavior.

The concept of "legacy DNA" in business operations is documented in Clayton Christensen's research on disruptive innovation, particularly in The Innovator's Dilemma (Harvard Business Review Press, 1997).

ABOUT THE AUTHOR

Susan Frew, C.S.P., didn't become an AI expert by studying algorithms in a lab. She became one by nearly losing everything and having to rebuild smarter.

As a former international VP with AT&T Wireless, Susan helped roll out mobile data networks when everyone said they'd never work. She led operations across two Caribbean countries, managed major restructures, and saw firsthand how technology adoption separates companies that thrive from those that fade.

After leaving corporate life, she built and scaled a multimillion-dollar HVAC and plumbing company that landed on the INC 5000 list. Then it nearly broke her. Inefficient processes, overwhelming operations, and her husband's health crisis forced a painful truth: they had to get out.

AI gave them a dignified exit. By systematically automating finance, operations, and marketing workflows, Susan made the business sellable instead of just closable. That experience, combined with over 3,000 hours of AI training and Google AI certification, transformed her from a struggling operator into one of the most sought-after AI implementation experts in the country.

Today, Susan delivers high-energy keynotes and workshops for Fortune 500 companies, associations, and industry conferences. Her mission: help business leaders use AI to reclaim time, cut waste, and build companies that serve their lives instead of consuming them.

She lives in Commerce City, Colorado, with her husband, William, and their dogs. Connect with her at susanfrewspeaks.com or on LinkedIn.